Y0-AAA-895

Musical Acoustics

Musical Acoustics

CHARLES A. CULVER, Ph.D.
Formerly Head of the Department of Physics, Carleton College

FOURTH EDITION

1956 **McGRAW-HILL BOOK COMPANY**
New York Toronto London

 Library of Congress Catalog Card Number 56-6953

15 16 MAMB 7 6 5

14904

To My Wife

Preface

The wide acceptance of *Musical Acoustics* during the past fourteen years has encouraged the author and the publishers to bring out a new edition of the book. This edition embodies many of the constructive suggestions made by teachers, and by those interested in music, who have made use of the earlier editions. A large part of the text has been more or less completely rewritten, and some rearrangement of the material has been made. At the suggestion of a number of teachers who have used the book as a text, several new features have been added, notably a chapter on basic concepts, and one on simple harmonic motion. Another addition has been the inclusion of a list of questions and problems at the end of many of the chapters. The Bibliography has also been expanded. In the main, the terminology of the American Standards Association has been followed. Except in a few cases where the vector concept is necessarily involved, the term "speed," rather than "velocity," has been used. An effort has been made to incorporate the latest findings of research workers in this field, including certain recent results secured by the author.

The helpful cooperation of those who have used the previous editions is deeply appreciated. The author is particularly indebted to Prof. C. M. Jensen, of Augustana College, who placed at his disposal a list of problems and questions that he has used in his classes in this subject. Many of the problems to be found at the end of the chapters are taken from that list. The cooperation of several manufacturers of acoustical equipment in supplying photographs of new equipment, and of a number of publishers in granting permission to reproduce certain illustrative material, is gratefully acknowledged. The cooperation of the many musicians who have assisted in the securing of technical data is deeply appreciated. The author also desires to recognize the valuable assistance of his wife in connection with the typographical and editorial aspects of the project.

This book is not a treatise on sound, but, as in the earlier editions, is intended to serve as a text for use by music majors in colleges and universities. However, in preparing the new edition, the author has kept in mind the interest of professional musicians and others who enjoy music. Music is both an art and a science. It is the author's belief that a knowledge of the means and processes by which music is produced, recorded, and transmitted will serve to contribute to the greater enjoyment of this, the noblest of the arts.

The author will continue to welcome constructive suggestions.

C. A. CULVER

Contents

Glossary

SYMBOLS

Δ and *o*	Used in connection with sound spectra charts to indicate a *trace*, and the *absence*, of some particular harmonic.
λ	The Greek letter *lambda;* used to indicate the length of a sound wave.
ω	The Greek letter *omega;* used to represent the product of 2π and the frequency.
/	Per.
f	Frequency; the number of vibrations per second.
s	Speed.
T	Period; the length of time required to describe a single vibration.
v	Velocity.

ABBREVIATIONS

C	Centigrade scale of temperature.
cm	Centimeter; one one-hundredth part of a meter.
cps	Cycles per second.
cu cm	Cubic centimeter.
db	Decibel; the unit of sound intensity.
F	Fahrenheit scale of temperature.
kc	Kilocycle; a thousand cycles.
log	Logarithm.
mm	Millimeter; one-tenth of a centimeter.
psi	Pounds per square inch.

SHM	Simple harmonic motion.
sin	An abbreviation for *sine;* a term denoting the numerical relation between the side opposite the angle and the hypotenuse in a right-angle triangle.
sq cm	Square centimeter.
vps	Vibrations per second.

TERMS

Antinodes	Points, lines, or surfaces in a transmitting medium at or on which the particles of the medium have maximum motion.
Complex tone	A musical sound containing two or more partials.
Dyne	A unit of force.
Erg	A unit of energy; also a unit of work.
Fundamental	The sound having the lowest frequency which a vibrating body is capable of emitting.
Half tone	Approximately one-twelfth of an octave.
Harmonic	The fundamental, or any upper partial whose frequency is a simple multiple of the fundamental frequency.
Microwatt	A millionth of a watt.
Nodes	Points, lines, or surfaces in a transmitting medium at or on which the particles constituting the medium are at rest.
Overtone	Synonym of *upper partial*.
Partial	Any component of a single complex tone.
Phon	A unit of subjective loudness level.
Pure tone	A single musical sound which is devoid of overtones.
Ratio	The quotient of the numerical value of one quantity when divided by the corresponding value of another quantity.
Sabin	A unit of sound absorption.
Tone	A general term used to designate any given musical sound emitted by a sonorous body; also used to indicate one of two principal *steps* in a musical scale.

Upper partial	Any partial having a frequency higher than that of the fundamental.
Watt	A unit of power.
Wave train	A succession of sound waves moving in some particular direction.
Whole tone	Approximately one-sixth of an octave.

1 *Basic Concepts*

1–1. *Introduction*

The glorious sounds that go to make up a symphonic composition involve certain fixed physical laws—laws that are commonly dealt with in the realm of physics. If we are to understand the physical basis of musical sounds it is therefore important and necessary to become acquainted with certain of these laws and concepts. For this reason we shall, in the sections that follow, give consideration to the nature of several fundamental relations that have to do with the elements of physics.

1–2. *Inertia and Mass*

As a beginning, let us consider the concept of inertia. The term refers to a general property of matter. Now inertia may be thought of as that property of matter by virtue of which it opposes any attempt to bring about a change in the velocity of a body. For instance, it is inertia that makes it necessary to apply a force to start or to stop an automobile. Inertia is sometimes loosely defined as the "laziness of matter." This resistance to a change of motion is the most fundamental and significant characteristic of matter. The magnitude of inertia is expressed in terms of what we refer to as mass; mass is a measure of inertia. For our purposes we may think of the terms mass and inertia as being synonymous. The unit in which mass is expressed is known as the gram. In the study of sound we shall be dealing with molecules of gases and solids. Such entities have very small masses, expressed in fractions of a gram.

The term weight is frequently confused with mass. These two expressions do not mean the same thing. Weight is the magnitude of the force with which the earth attracts a body. Experience shows that the weight of a body is proportional to the mass. However, the weight of a body depends upon its distance from the center of the earth, while the mass (or inertia) of a body is the same regardless of its physical environment. For example, our weight is less on a mountain top than it is at sea level, but the mass of our physical body is the same whether we are on the earth or on the moon. As we proceed with our study of musical sounds we shall encounter the concept of mass. The reader should therefore acquire a clear idea of the significance of this important term.

1–3. *Displacement*

Let us next consider the concept of displacement. If a body is moved from one point to another we say that displacement has taken place. If the displacement is along a straight line we are dealing with linear motion, and the magnitude of the displacement involved may be expressed in any unit of length, such as inches or centimeters. In considering the matter of displacement it frequently becomes necessary to specify the direction of the motion, for instance whether the displacement is up or down, or to the right or left, of some reference point or line. Displacement, therefore, involves both magnitude (greatness) and direction. The concept of displacement will enter into the consideration of those motions having to do with the generation and transmission of sounds.

1–4. *Speed and Velocity*

When linear motion takes place we are usually interested in the rate of displacement. For instance, if we walk eight miles in two hours, our rate of walking is four miles per hour. This time rate of displacement we refer to as speed—in this particular case, the average speed. The speedometer on our car indicates not the average speed but the instantaneous speed—the speed at any given instant. Under certain circumstances we may be interested not only in the speed but also in the direction of the motion. If one were to say that we were traveling at 30 mph in a northerly direction we would be specifying not only the speed (magnitude factor) but also the direction aspect of the displacement. In short, we would be dealing with what may be referred to as velocity. It will thus be seen that the terms speed and velocity are not synonymous; speed involves magnitude (greatness) only, while velocity has to do with both magnitude and direction. Two bodies may be traveling with the same speed but if they are not moving in the same direction they do not have the same velocity. The two terms, speed and velocity, are often, though incorrectly, used interchangeably. In dealing with musical sounds we shall have occasion to employ both concepts, but the distinction should always be kept clearly in mind. Speed is expressed in feet per second or centimeters per second.

In the foregoing brief discussion concerning speed and velocity we have tacitly assumed that we are dealing with motion in a straight line, that is, rectilinear displacement. There are, of course, many other types of motion, such, for instance, as the path taken by a jet of water from a hose, but for our purpose, straight-line (rectilinear) motion will be the type ordinarily encountered.

1–5. *Acceleration*

Another assumption that we have thus far made is that the velocity of the motion being considered is uniform; that is, that neither the speed nor the direction changes. Actually, however, either or both of these factors may change while the motion is taking place. For instance, when we are driving a car we may increase or decrease the speed, and we may also change the direction of the motion. If either the speed (magnitude factor) or the direction aspect of the motion is changed, we say that acceleration has taken place. A car moving around a curve at a fixed speed of, say, 30 mph would not have uniform velocity; it is being accelerated. Acceleration, then, may be thought of as the time rate at which velocity changes, and this, we repeat, may have to do with either the speed or the direction factor, or both, simultaneously. If we are traveling on a straight road at a constant speed of, say, 50 mph, the acceleration is zero. If we go around a curve at the same constant speed, the acceleration is not zero.

Now it is necessary in this connection to recognize the fact that this change in velocity does not necessarily take place at a constant rate; that is, the acceleration may not be uniform. For instance, we may increase (or decrease)[1] the speed of a car at the rate of 10 mph for 5 sec and then change its speed at the rate of 15 mph in the next 5 sec, and so on. But this is a long story and does not concern us at present. We shall later be chiefly concerned with maximum and minimum values of velocity and acceleration, and in the direction aspects of these two factors. In passing it is to be noted that acceleration is expressed in ft/sec/sec or in cm/sec/sec.

1–6. *Force*

Thus far we have dealt with motion as such without referring to the cause of the displacement involved. Let us now consider the factor that gives rise to, or modifies, the motion of a body. The term force is given to this causative factor. We are all familiar with the fact that muscular force is required to move a body which is at rest or to increase or decrease the rate at which it is moving when it is once in motion. Force is also required to change the relative position of the parts of a body, as when we stretch a rubber band or compress a body of gas. We may define force as that which tends to produce a change in the motion of a body or to change the relative position of its parts.

There are many types of force, the most common being the pull exerted by the earth on a body. This force we call weight. Friction, gas pressure, the pull exerted by a magnet, are other common examples of force. In our study of musical sounds we shall be chiefly concerned with the action of forces that tend to bring about acceleration (a change in velocity). What is probably the most important relation in the whole realm of physics has to do with force and acceleration. It has been found that the magnitude of the force necessary to bring about a change in the velocity of a body depends upon the mass (inertia) of the body; it is also known that the acceleration produced is proportional to the force applied. One may therefore say that

$$\text{Force} = \text{mass} \times \text{acceleration}$$

[1] A decrease in speed is referred to as negative acceleration, or "deceleration."

The magnitude of a force may be expressed in various units, but the unit most commonly used in the study of sound is the dyne, a word from the Greek meaning force. Now the dyne represents a very small force. A gram weight is pulled by the earth with a force of nearly 1000 dynes. The atmosphere exerts a pressure of something like 14.5 pounds per square inch. This is equivalent to about a million dynes per square centimeter. Sound waves, as we shall see, exert pressures of the order of 0.0002 dyne/sq cm, in the case of faint sound, and 2 or 3 dynes/sq cm for loud sounds.

1–7. *Elasticity*

In beginning our discussion of useful concepts we referred to a general property of all matter, viz., inertia. We will find it useful to give some thought to a special property possessed by certain materials. This property is known as elasticity, and, as we shall see, this particular characteristic plays a very important part in the generation and transmission of sound. It therefore is important that we become acquainted with the nature of this special characteristic of certain substances.

It is a matter of common experience that certain bodies, such as a strip of wood or a body of gas, when subjected to a force tending to cause a deformation, will, upon removal of the applied force, return to their original shape and size. Bodies that do recover from such distortion are said to be elastic, and materials, such as putty or wax, that do not react in this way are said to be inelastic. Elasticity, then, might be defined as that property by virtue of which certain kinds of material tend to recover their original shape or size after undergoing deformation. As we proceed we shall be particularly interested in the elasticity of wood, steel, and air.

1–8. *Work*

There are one or two other concepts that we shall encounter in the study of sound. One of them is the idea of what is referred to as work. Whenever the application of force results in the movement of a body as a whole, or the relative movement of its parts, work is done. Work is done when water is pumped from a well, when a body is dragged along the ground in opposition to a frictional force, when a gas is compressed, when a spring is stretched or compressed, when a storage battery is charged. A man holding a pail of water in his hand is exerting a force in opposition to the pull of the earth, but he is doing no work in the mechanical sense of the term. Displacement of the body as a whole or the relative displacement of its parts in the direction of the applied force is an essential condition. Common experience tells us that one does twice as much work in walking up two flights of stairs as is done in going up only one flight. Since, then, force and displacement are both factors in work, we may say that the numerical measure of work is given by the product of the applied force and the effective displacement. Thus

$$\text{Work} = \text{force} \times \text{displacement}$$

By "effective displacement" is meant the displacement in the direction of the applied force.

The unit of work is the erg; it is the amount of work done by a force of one dyne in overcoming a resistance through a distance of 1 cm. Since the dyne is a very small unit of force, the erg represents a correspondingly small amount of work. Since the earth exerts a pull of about 980 dynes on a gram weight, if we lift such a weight through a distance of one centimeter against the force of gravity, we do nearly a thousand ergs of work.

1–9. *Energy*

An elevated weight, a stretched spring, a compressed gas, a charged storage battery, a moving bullet, and a rotating flywheel are all capable of doing work. We say that they possess energy. Work has been done upon them to bring about the state of stress or motion in the cases cited. A body or mechanical system possesses energy when and if, by virtue of its position, the relative position of its parts, or its motion, it is capable of doing work. The energy that a body or system possesses is measured by the work it can do. In order to do work this energy must be expended. Because of that fact, the unit in which energy is expressed is naturally the same as that used for work, viz., the erg. The elevated weight, the stretched spring, the compressed gas—these bodies possess energy by virtue of their position or configuration. Bodies in which energy resides because of the conditions just enumerated are said to possess potential energy. The energy that a body possesses by virtue of its motion (the moving bullet and the rotating flywheel) is known as kinetic energy. In our later study we shall be concerned chiefly with energy of the kinetic type.

1–10. *Power*

Thus far in our discussion of work we have made no reference to the length of time that may be consumed in accomplishing a given amount of work. In appraising the final effect due to the expenditure of energy, it is more or less evident that the time element will enter as a factor. If we have a field to plow or a house to build, we are interested not only in the total amount of work involved but also in the length of time required to accomplish that particular piece of work. The factors of work and time are both involved in the concept of power. This relation may be expressed thus:

$$\text{Power} = \frac{\text{work done}}{\text{time involved}}$$

From the foregoing it will be evident that the term power signifies the rate of doing work—the rate at which energy is expended. It is thus to be noted that power will be expressed in ergs per second. Power does not mean strength or force or work. The reader should carefully note the distinction. Since, as we have seen, the erg is a very small unit of work, it has been found convenient to establish a larger unit of power known as the watt, which is the equivalent of 10^7 ergs/sec. When energy is expended at the rate of 10^7 ergs/sec, the power is 1 watt. In a small electric light bulb the rate of energy expenditure may be, say, 25 watts. In dealing with sound we shall find that the power involved will be relatively small—of the order of a fraction of a watt.

QUESTIONS

1. Cite an example that will illustrate the significance of the concept of inertia.

2. How would you define *displacement?*

3. What is the real distinction

between the concepts of speed and velocity? Cite examples to support your statement.

4. What are the units in which speed is expressed?

5. What do you understand by the term *acceleration?*

6. What does the term *force* imply, and why is this concept important in the study of sound?

7. Cite several substances that exhibit the property of elasticity; also mention some inelastic substances.

8. In what way does the property of elasticity enter into the study of sound?

9. If a person holds a pail of water in his hand, is he doing work? Why?

10. In what units is energy expressed? Power?

11. Distinguish between energy and power.

12. What relation exists between work, energy, and power?

2 *Simple Harmonic Motion*

2–1. *Simple Harmonic Motion*

Having examined certain important concepts that we shall encounter in our study of musical sounds, it will now be advisable to consider the type of motion involved in the generation and propagation of sounds in general.

In dealing with mechanical, electrical, and acoustical phenomena one encounters various types of motion. Reference has already been made to motion in a straight line (rectilinear motion). We are all more or less familiar with the motion of the water particles forming the spray from a garden hose. Such particles trace a curved path, and we refer to such a type of movement as curvilinear motion. Another form of curvilinear motion is described by a reference mark on a revolving wheel. Here we find that the point in question moves in a circular path, and, what is still more significant, the motion repeats itself in equal intervals of time. The point on the wheel is also describing curvilinear motion, but it differs from the water jet in that its motion is periodic.

Now there are many types of periodic motion, as, for instance, the to-and-fro motion of a clock pendulum, the movement of the balance wheel of a watch, the up-and-down motion of a weight suspended from a coiled spring, the to-and-fro motion of a violin string, or the up-and-down motion of the particles constituting a water wave.

If, in the simple experiment illustrated in Fig. 2–1*a*, the motion is initiated by pulling aside the free end of a wooden strip and then releasing it, it will be observed that the vibratory motion repeats itself, gradually ceasing with time. The same general type of phenomenon may be observed if we pull down a weight suspended as shown in Fig. 2–1*b*. The question at once arises: Why does the motion continue to repeat itself? This is a very important question. In both of the cases cited, it will be noted that after the body was released it tended to return to its resting position and that it also overshot that initial position. This occurred because in both instances the deformation of the strip, or the spring, gave rise to a restoring force. This restoring force is due to the elasticity exhibited by the wood and the coiled spring, and its magnitude depends upon the displacement from the rest position, and this is significant. The restoring force caused the body to return to its original position. But why does it overshoot the original position? We have already learned that all bodies have inertia and that because of this they tend to continue in a state of motion or rest. In our examples the bodies have inertia, and because of this they tend to continue the motion initiated by the restoring force, and hence they overshoot the median position. Thus the strip and the weight describe a periodic motion, and such a type of motion is referred to as simple periodic motion, or, if you please, simple harmonic motion. One encounters such a type of motion in dealing with the study of electricity, light, and sound. In fact, simple harmonic motion (abbreviated SHM) is, perhaps, the most important and significant type of motion in all nature. And it should be noted in passing that such a form of motion, in the case of sound, always involves the properties of elasticity and inertia.

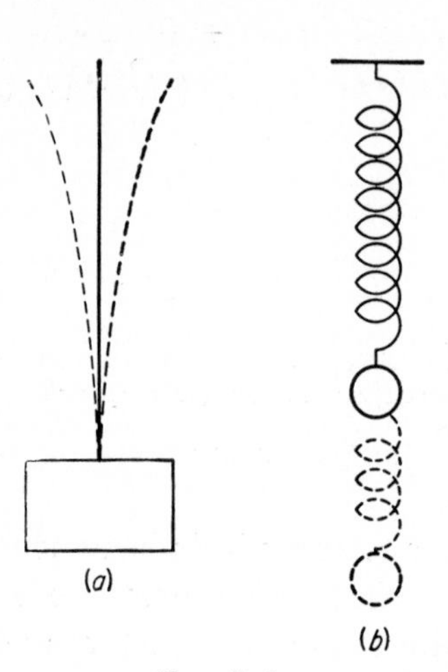

Fig. 2–1

2–2. *Properties of SHM*

Now this type of periodic motion is unique in several respects. The characteristics of SHM should be carefully noted. In the first place, as we have already seen, the magnitude of the restoring force at any instant is proportional to the displacement at that particular instant. The important characteristics of SHM may be readily observed by means of a simple experiment. If a small metal or wooden ball is suspended from the ceiling by means of a light thread and caused to move, by means of an initial impulse, in a circular path, while it is illuminated by a more or less parallel beam of light, the shadow of the ball on a plane, as it describes circular motion, will simulate simple harmonic motion. (There is an intimate relation between uniform circular motion and SHM, but it would not serve our purpose to investigate this relation here.)

If one views the moving shadow of such a conical pendulum in the plane of the circular motion, it will be noted that the motion of the shadow possesses several striking and significant characteristics. In the first place its linear speed is greatest when passing its median position c, as indicated in Fig. 2–2. At each extremity of its oscillation (p_1 and p_2) its speed is zero. It also is to be noted that the motion is linear. At some particular instant, say when the shadow is at the point p_3, the displacement has a value of x, as measured from its median position. It is customary to think of the displacement to the right of the median position as positive, and that to the left of c as negative. While it is not apparent from the experiment, it may readily be shown that the acceleration is a maximum when the displacement is a maximum, as at p_1 and p_2, and that the speed is a maximum when the displacement is zero, viz., at the median point. Remembering our definition of velocity, it will be evident to us that, while the speed at p_3 is the same as it is at a corresponding point to the left of c, the velocity is not the same at those two points in the path: the motion of the object is oppositely directed at these two points. What has been said about the shadow holds also in the case of the vibrating stick or the oscillating weight or the coiled spring, previously cited.

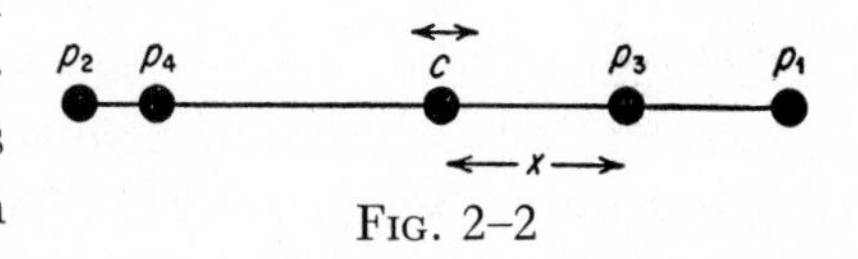

Fig. 2–2

2–3. *Graphic Representation of SHM*

It often serves a useful purpose to set up a graphic representation of SHM. A simple method by which this may be done is illustrated by the experiment depicted in Fig. 2–3. If a swinging pendulum be arranged to make a trace on a surface moving at constant speed, as indicated, the curve traced will be a graphic representation of the SHM described by the oscillating pendulum. Methods are available whereby a graphic record of any type of SHM may be secured. Fig. 2–4 shows an enlarged sketch of this form of record. It represents the displacement when plotted against time. Because the shape of such a curve can be represented by a certain mathematical equation it is commonly referred to as a sine curve, and the periodic motion represented by such a graph is said to be sinusoidal in character. The variations in velocity and acceleration in SHM are also sinusoidal in character, as shown in Fig. 2–5.

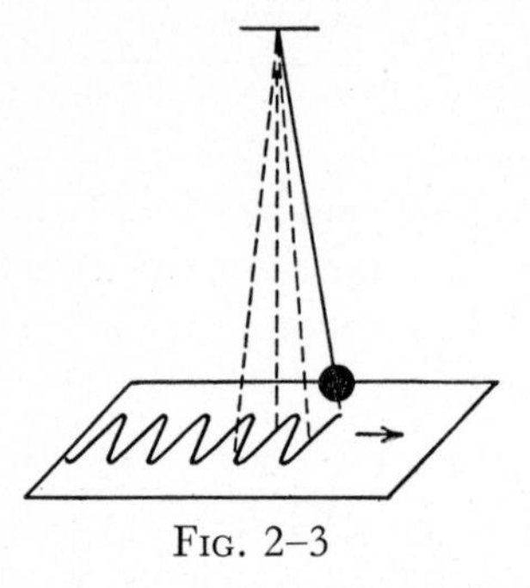
FIG. 2–3

Again referring to Fig. 2–2, the maximum displacement cp_1 or cp_2 is what is referred to as the amplitude of the SHM. This quantity is also represented by a or a' in Fig. 2–4. The length of time

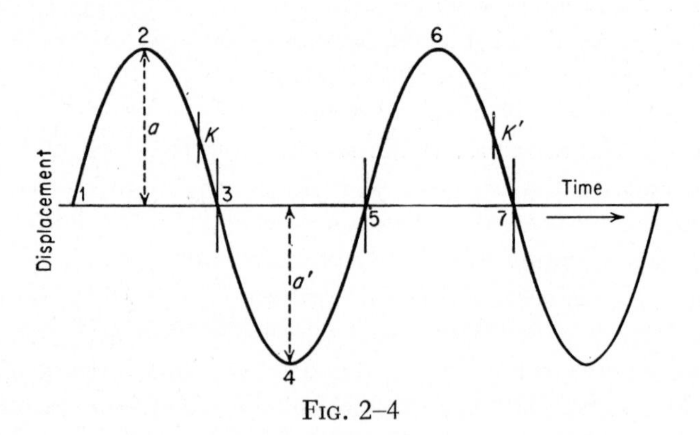

FIG. 2–4

required for the body or particle to describe a complete "round trip," say (in Fig. 2–2) from c to p_1 to c to p_2 and to c again, is known as the period of oscillation; or, to put it another way, it is the time required for the body to describe a complete cycle, as shown in Fig. 2–5. The number of complete excursions described per second is spoken of as the frequency of the oscillations.

One other important term used in the discussion of SHM should be noted, and it stands for a very important concept. Reference is here made to the term phase. In general terms, phase may be thought of as indicating the position and direction of motion of an oscillating body or particle with reference to some specified point in its path. Phase is commonly expressed as a fraction of a period. To be specific, it is the fraction of a period that has elapsed since the moving body last passed the median point (in Fig. 2–2) when moving in a positive (to the right) direction. Referring to Fig. 2–2, if we take c as the point of reference, and assume that the body is at p_3 and moving to the right, its phase would be the time that had elapsed since it passed from c to its present position. On reaching p_1 its phase would be $T/4$, where T stands for the period. If the particle stopped at p_1 (as it would) and then moved back to c, its phase would be $T/2$ or half a period. If the body continued to move to the left, after it reached c and arrived at the point p_4, its phase would be the time required for it to move from c to p_1 and back to p_4. In the diagram shown in Fig. 2–4, any two points, such as k and k', are in like phase.

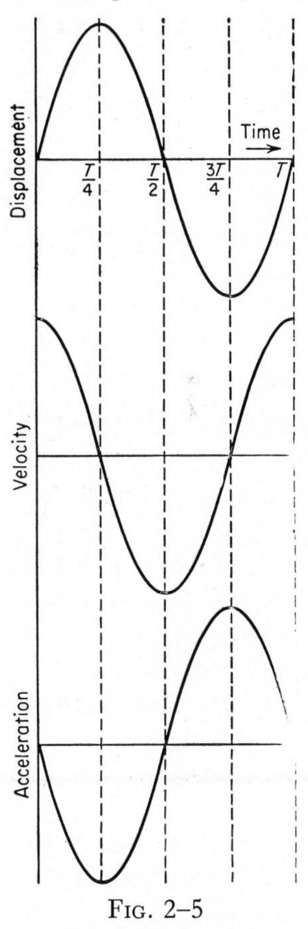

FIG. 2–5

The foregoing discussion leads to the concomitant concept of phase difference. The phase difference between points 1 and 2 (Fig. 2–4) is one-fourth period; between 1 and 3 a half-period, etc. Both phase and phase difference play important parts in the production and transmission of sound, as we shall see later.

As applied in the study of sound, one might summarize the characteristics of SHM as follows: it takes place in a straight line; it repeats itself in equal intervals of time; there are instants of rest at the two extreme positions; the maximum speed occurs at the median point of its oscillation; the period of vibration does not depend upon the amplitude; after the exciting force ceases, the magnitude of the restoring force at any point in its swing depends upon the distance of that point from the median point.

2–4. *Simple Waveforms*

Since the shape of the graph in Fig. 2–4 is wavelike in appearance, the curve is said to show the waveform of the oscillations involved. Now any such waveform may possibly differ in several important respects from a waveform representing some other periodic motion. For instance, two oscillations may have the same frequency but differ in amplitude, as indicated in Fig. 2–6*a*, the amplitude of one, in this case, being twice that of the other. Or the two simple periodic vibrations may have the same amplitude but differ in frequency, as shown in (*b*) of the same figure. All of these oscillations are of the same general form, but, as seen, they differ in two important respects: the graphic representations of the several variations make these differences apparent.

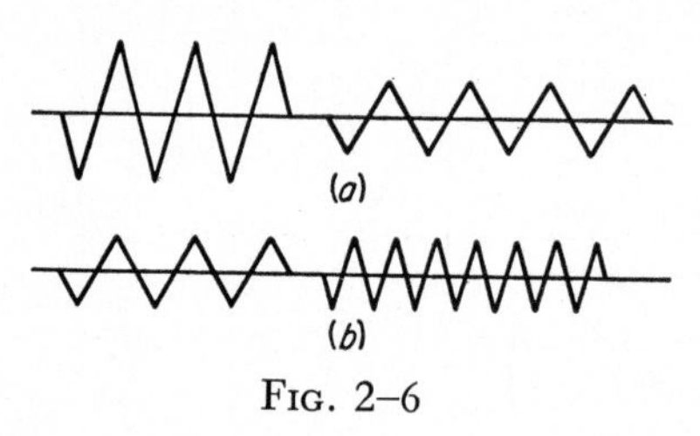

FIG. 2–6

2–5. *Simultaneous Vibrations*

It is an interesting fact connected with mechanics that a given body may be caused to vibrate in more than one SHM at one and the same time. In other words, simultaneous vibrations may take place. In such a case the motion is not simple, but is made up of a more or less complicated type of vibration. But there is a remarkable and highly significant interrelation between the several modes of vibration constituting such a complex motion. In 1622 Fourier, the great French mathematical analyst, showed that it is possible to break down, analytically, any complex periodic motion into a series of simple motions and that each component in the series is a simple periodic motion. Furthermore, he pointed out that the frequencies of the several component motions have values that are twice, thrice, etc., that of the lowest frequency present. This is a truly remarkable relation and it is readily applicable, as we shall see later, to the periodic motions that constitute sound. We have seen that a simple periodic motion may be graphically represented by a symmetrical curve, as shown by *A* or *B* in Fig. 2–7. For example, let us say that the frequency in *B* is twice that in *A*. Either of these curves may be represented mathematically by the simple equation

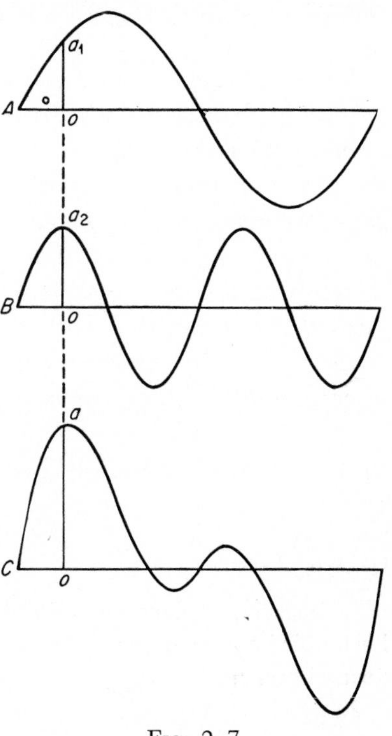

Fig. 2–7

$$a_1 = k_1 \sin \omega t$$

where $\omega t = 2\pi f \cdot t$

a_1 the instantaneous displacement of any given particle from some reference line; k_1 the maximum displacement or amplitude; f the frequency of the oscillation; and t the elapsed time since the particle passed its point of zero displacement. The term $(2\pi f \cdot t)$ indicates that the position of the particle in its cycle of movement is a periodic function of the time. If we let the above equation represent the periodic motion shown by *A*, the corresponding expression for *B* would take the form

$$a_2 = k_2 \sin 2(2\pi f \cdot t)$$

since, in this case, the frequency in *B* is twice that of *A*. If one simultaneously impresses upon any given body, or part thereof, the two motions represented by *A* and *B*, the resultant motion will be represented by the graph *C*, and this resultant curve will be given by an expression having the form[1]

$$a_1 + a_2 = a = k_1 \sin (2\pi f \cdot t) + k_2 \sin 2(2\pi f \cdot t)$$

[1] Those who are acquainted with calculus will be interested in reading several paragraphs on Fourier's theorem to be found in Barton's "Textbook of Sound,"

Actually we are not confined to two components. In our subsequent study of musical sounds we shall encounter cases in which a body may vibrate simultaneously in as many as 20 modes. The resultant vibration can, in any such case, be represented mathematically, as in the simple case cited above.

The resultant waveform, as given by curve *C*, Fig. 2–7, may also be obtained by a graphic method. If we examine the situation at some particular instant, as indicated by the vertical line, and if at the time indicated we add the value of the displacement oa_1 to that at oa_2 we get oa, which is the corresponding displacement of the resultant motion at that instant. If we do this at equal time intervals throughout a complete cycle we can, by connecting the points thus located, draw a graph which will represent the waveform of the resultant motion. And this procedure is applicable to any number of components. Later, when we come to study the musical sounds emitted by the various musical instruments, we shall have occasion to deal with many complicated waveforms.

QUESTIONS

1. In what respect does simple harmonic motion differ from uniform circular motion?
2. Cite several examples of simple harmonic motion; of two or three nonharmonic motions.
3. What is the significance of the term *simple* in connection with harmonic motion?
4. Account for the fact that, in the case of simple harmonic motion, the velocity is zero when the displacement is a maximum.
5. At what part of a cycle is the velocity greatest?
6. What is meant by the term *period? Frequency?*
7. Draw a simple harmonic curve, and then sketch in another like curve representing a motion that differs in phase by one-fourth period from the first.
8. What does the term *wavelength* mean?
9. What are the distinguishing characteristics of simple harmonic motion?
10. What is the significance of the term *waveform?*
11. Define the term *sinusoidal.*

---

pp. 83ff. Professor Miller in his lectures on "The Science of Musical Sounds" shows how this theorem is applied to problems connected with sound analysis; see pp. 92ff.

12. Draw a curve representing a periodic motion having a definite amplitude and wavelength. On the same sheet draw the curve of a motion whose period is one-half that of the first, and whose amplitude is one-third that of the first. Let the second motion differ in phase by one-eighth period from the first. Now sketch in the resultant waveform.

13. In representing a longitudinal wave motion by a moving sine curve, what points on the curve represent the points of (*a*) maximum compression; (*b*) maximum rarefaction; (*c*) normal pressure?

3 *Nature and Transmission of Sound*

3–1. *Nature of Sound*

Having examined certain basic concepts and relationships having to do with motion in general, we are now in a position to make use of these fundamental ideas as we proceed to consider the nature of sound and the laws that govern its propagation through space.

Before proceeding with such a study, it will be well to arrive at a clear understanding of what is meant by the term *sound.* For our purposes it may be said that **sound is any vibratory disturbance in a material medium which is capable of producing an auditory sensation in the normal ear.** We are here dealing with the objective **cause** of a sensation, and **not** with the sensation itself. All such disturbances have their genesis in some body which is undergoing vibratory motion. Such a sonorous body may be a solid, a liquid, or a gas. The vibratory motion of the sounding body is usually conveyed to the auditory receptor, the ear, by means of a wave motion in a gas or mixture of gases such as the air, though solids and even liquids may take part in such a transfer.

The laws which govern the transfer of energy from the sonorous body to the ear will be considered in detail later. For the present we need only consider what is meant by two or three other terms. The term *wave motion* was used above. This term is employed to designate any periodically recurring disturbance in any medium through which energy is being propagated. Everyone is familiar with ripples on the surface of water. In that case the particles of the medium move up and down at right angles to the direction in which the waveform is traveling. Such a disturbance is referrred to as a **transverse** wave. In the case of sound the particles constituting the medium, instead of moving transversely to the direction in which the wave is traveling, execute a to-and-fro (periodic) motion **in the direction in which the disturbance is being propagated.** A disturbance of this character is spoken of as a **longitudinal wave.** The particles, then, which constitute a medium through which a sound wave is traveling are describing longitudinal periodic motions. A sonorous body communicates its vibratory motion to the adjacent air particles, and these in turn bring about a corresponding motion of the neighboring particles. Motion, and the accompanying energy, is thus conveyed from particle to particle. Thus the energy of the sonorous body is transferred to the receiving agent, viz., the mechanism of the ear. This will be better understood by referring to Fig. 3–1.

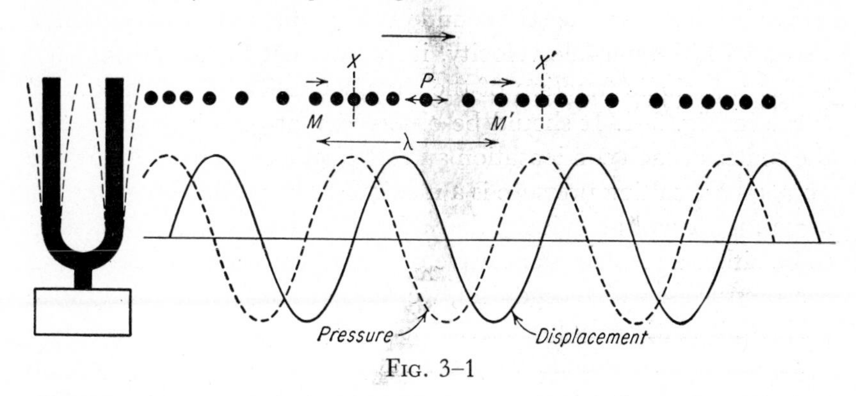

FIG. 3–1

If an ordinary tuning fork, for example, be struck gently, its prongs will be set into vibration, as indicated by the dotted lines in the sketch. This vibratory motion will give rise to a series of condensations and rarefactions in the surrounding medium. That is, the movement of a given prong of the fork, say to the right, will cause the particles which are near it to become crowded together; as the prong swings toward the left it tends to create a partial vacuum behind it. It will thus be seen that, because of the vibratory motion of the fork, the density of the medium will become alternately greater and less than normal (see upper part of the sketch).

Thus the particles of air are caused to describe a to-and-fro (a periodic) motion, and this to-and-fro displacement of the particles of the medium constitutes the sound wave. In the diagram, the black dots represent the air particles; one condensation region is shown at the left of the row, corresponding to a rightward swing of the fork. The distance through which the particles move is extremely small, being of the order of a small fraction of a millimeter. It is also to be noted, in this connection, that this movement of the particles constituting the medium gives rise to a pressure change, called the **sound pressure:** the magnitude of this pressure change depends upon the velocity of the particles involved. Remembering what was said in Sec. 2–2, it is evident that we have here a case of simple harmonic motion. In our earlier discussion of SHM it was noted that the displacement, and the particle velocity, in such a type of motion is a periodic function of the time, i.e., the instantaneous value of the displacement, or velocity, varies with the elapsed time. Now since the sound pressure due to the wave motion varies with the particle velocity, it follows that the sound pressure is also a periodic function of the time, as indicated by the dotted curve in Fig. 3–1. It should be observed that the displacement at the center of each condensation and rarefaction is zero. But at these points the resulting pressure is a maximum. From the foregoing observations it will be evident that a sound wave may be considered to be either a displacement wave or a pressure wave. Since the ear is a pressure-sensitive organism, we shall be concerned chiefly with periodic variations in pressure.

Figure 3–1 should be very carefully studied, particularly with reference to the concepts of wavelength and phase. The particles M and M' are in like phase with respect to the reference planes x and x' respectively. Each particle has consumed equal fractions of a complete period in moving from the reference plane to its present position.

Figure 3–1 also indicates what is meant by the term wavelength. This is an expression that we shall frequently encounter in our study of musical sounds. The wavelength of any progressing periodic disturbance is the least distance, measured along the path of progression, between two points in like phase, or state of motion. In the diagram the distance designated by the Greek letter lambda (λ), between any two particles such as M and M' (that are in like phase), is a measure of the wavelength involved in that particular wave disturbance.

Previously we have defined the terms speed and frequency. We are now in a position to note an important relation between these two factors and wavelength. It may be shown that

$$\text{Wavelength} = \frac{\text{speed}}{\text{frequency}}$$

or

$$\lambda = \frac{s}{f}$$

If speed (s)* is expressed in cm/sec and frequency (f) in cycles/sec, the wavelength (λ) will be given in cm. If s is expressed in ft/sec, λ will be in ft. For instance, let us assume for the moment that the speed of sound in air is 1100 ft/sec, and that middle C on the piano has a frequency of 262. Substitution in the above relation then gives

$$\lambda = {}^{1100}\!/_{262} = 4.2 \text{ ft, approximately}$$

In other words, the wavelength of the note we recognize as middle C is about 4 ft. On the same basis the note two octaves above middle C would have a wavelength of slightly over 1 ft.

One final word of caution in connection with the use of graphic methods of representing wave motion: It should always be remembered that the undulatory lines used to represent the periodic changes in displacement or pressure constitute a convenient way of representing the periodic disturbances occurring in the medium through which the wave motion is being propagated. Such curves do not constitute a picture of the actual motion of the oscillating particles.

3–2. *Noises and Musical Sounds*

All sounds may be classified roughly into two groups, viz., noises and musical sounds. While there is no strict line of demarcation between these two types of sounds, a noise commonly consists of a group of nonperiodic pulses arising from the irregular vibration of

* Some writers use the terms *speed* and *velocity* interchangeably. In such cases the letter c is used to designate velocity.

a body or group of bodies. Such a sound does not manifest a definite pitch, and usually produces an unpleasant auditory sensation. On the other hand, a sound which we classify as musical gives rise to a pleasing sensation and is characterized by being periodic and having a definite pitch. Musical sounds are by no means simple, however; they may indeed be highly complex in character, as we shall see later. Furthermore, it should be observed that speech and musical sounds are often accompanied by a noise background. In fact, noise is sometimes defined, in broad terms, as any unwanted sound. We shall later have occasion to refer to this aspect of sound.

3–3. *Speed of Sound*

The speed with which a succession of sound waves travels through a given medium depends upon the nature and condition of the medium. The density of the medium and its degree of elasticity are the most significant factors in this connection. The relation between these two factors may be expressed thus:

$$s = \sqrt{\frac{k}{d}}$$

in which s is the speed of sound, d the density (mass/unit volume) of the medium, and k a constant depending for its value on the degree of elasticity of the medium involved. A change in temperature produces a change in both the elasticity factor and the density. These changes are, however, of such relative magnitude and direction that the net result is an increase in speed for an increase of temperature. In air at a temperature of 20°C the speed is 344 m/sec (1129 ft/sec). At zero degrees the velocity is only 331.7 m/sec (1088 ft/sec).[1] It has been determined that, in the case of air, the velocity changes 0.6 m/sec per degree C. (1.1 ft/sec/degree F). In wood the velocity is of the order of 4000 m/sec, and in steel about 5000 m/sec. Following is a short table giving the velocity of sound in a number of common media.

[1] The reports on research conducted to determine the velocity of sound give widely varying values, ranging from 330.6 to 331.9 m/sec. The phenomenon of dispersion probably accounts, in part, for the various values reported.

Substance	Temp., °C	Speed	
		m/sec	ft/sec
Air	0	351.7	1087.5
Air	15.6	341.0	1118.0
Hydrogen	0	1269.5	4165.0
Hydrogen	15.6	1363.1	4340.0
Oxygen	0	317.2	1041.0
Carbon dioxide	0	263.0	863.0
Water	15	1437.0	4714.0
Steel		5000.0	16400.0
Maple wood (along fiber)		4111.0	13444.0
Pine (along fiber)		3320.0	10889.6

An examination of the data given in the foregoing table reveals several important facts. In the first place, it will be noted that at ordinary room temperature sound waves will travel in air about 110 ft in $\frac{1}{10}$ sec. This should be kept in mind when we come to consider the matter of echoes and reverberation (Sec. 3–6). It may also be noted that sound travels at the rate of about 13 miles a minute, or 768 mph. When one recalls that light waves have a speed of approximately 186,000 mps, it will be obvious that the speed of sound is relatively small. Certain types of aircraft now travel at a speed greater than that of sound. That sound does travel at a relatively slow speed in air is apparent if one observes a carpenter at work on a building located several hundred feet away and listens to the sound resulting from the stroke of a hammer. It will be noted that the workman has lifted the hammer for a succeeding stroke before the sound of the last stroke reaches the listener.

Again referring to the foregoing table, it will be noted that the speed of sound in hydrogen gas is roughly four times greater than that in oxygen, at standard temperature and pressure. Now the elasticity of these two gases, at the same pressure, is approximately the same, but the mass of hydrogen molecules is only one sixteenth that of oxygen molecules. The density of oxygen at standard temperature and pressure is therefore sixteen times greater than that of hydrogen. In the equation giving the speed of sound we see that the density (d) is in the denominator of the fraction, and as a square root factor. A simple algebraic computation thus will show why the speed of sound in hydrogen is fourfold what it is in oxygen.

In dealing with solids, however, the case is somewhat different, and significantly so. If one compares the speed of compressional sound waves in steel to that in air it will be found that the speed in this particular metal is some fifteen times what it is in air. A similar comparison between air and maple wood shows a ratio of about 12.3 to 1 in favor of the wood. Now the density (d) for both steel and wood is greater than that of air, but the elasticity factor (k) is enormously greater in the case of the solids as compared to that of the gases. This accounts for decidedly greater speed in solid materials.

We shall see later that the speed of sound waves is an important factor in the design and functioning of various musical instruments, as well as in the problems connected with architectural acoustics. In any given homogeneous medium **the speed of all sound waves is the same.** In other words, the musical sounds made by a piccolo travel at the same speed as do the musical sounds evoked from a bass horn. Were this not so, orchestral renditions and choral singing would not be feasible.

3–4. *Refraction of Sound*

In a homogeneous medium a sound disturbance tends to move outward from the source in the form of a sphere, as indicated in the cross-sectional view of a sound wave shown in Fig. 3–2.[1] At relatively great distances from the source, the wavefront approaches that of a plane surface, and we have what is known as a *plane* wave.

If, however, the density or elasticity of the medium is not the same in all directions, because of temperature differences, for instance, the wavefront may be tilted, which means that the direction

[1] These remarkable photographs of sound waves were made by Prof. A. L. Foley, of Indiana University. The round object in the center is one of a pair of balls between which an electric spark is caused to jump. An electric spark also serves as the source of light by which the photograph is made. The compressional wave changes the optical density of air with the result that the photographic plate records what amounts to an image of the sound wave. This general technique has proven to be exceedingly useful in the study of acoustical problems, particularly those connected with acoustics of buildings.

of propagation will be changed. This phenomenon is known as refraction. Under such circumstances the sound ray (direction of propagation) is bent; that is, it undergoes refraction. During periods of low temperature for instance, where the ground is covered with snow, the speed of sound will be relatively low near the ground, with the result that the sound ray will be refracted downward, as shown in Fig. 3–3*a*. The greater part of the energy of the sound wave will thus be confined to the layer of air closely adjacent to the surface of the earth. Under these conditions, particularly if the air is quiescent, sounds can be heard at great distances. In general, in warm weather the temperature of the air decreases with altitude, and hence the speed of sound decreases with height, and the resulting refraction tends to tilt the wavefront, as shown in Fig. 3–3*b*. Under these conditions sounds cannot be heard at great distances. The difference in the temperature of the air near the ceiling of an auditorium as compared with that near the floor gives rise to refraction effects, and thus becomes a factor in architectural acoustics.

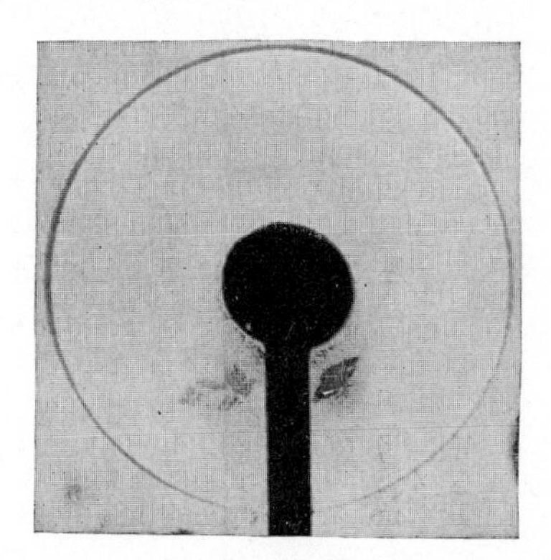

FIG. 3–2

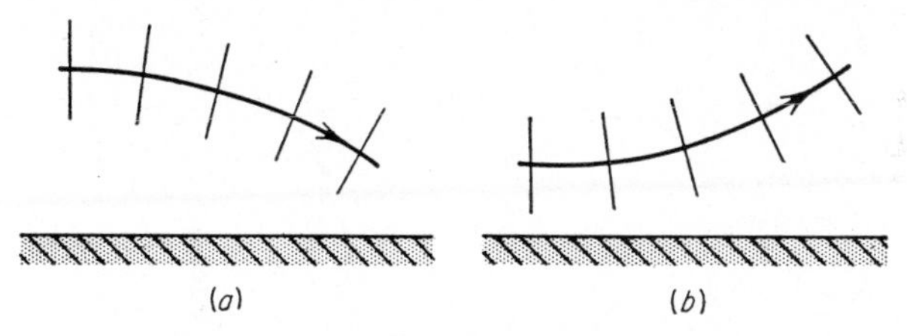

FIG. 3–3

In the open air, wind may give rise to refraction effects. In general, any wind has a greater velocity high above the earth than at the surface, because of friction with the earth's surface. (The term *velocity* has purposely been used here, rather than speed, because the direction factor is involved in this connection.) Suppose that a wind is blowing, say, 20 ft/sec in a southward direction at the earth's surface and at 50 ft/sec at some distance above the ground. Let us assume that a source of sound is located north of an observer. If, in still air, the speed of sound be taken to be 1100 ft/sec, the effective velocity of the sound near the ground in a southward direction will be 1100 + 20, or 1120 ft/sec. In the upper air the effective velocity will be 1100 + 50, or 1150 ft/sec. The wavefront will therefore be deflected (refracted) downward, with the result that, because of the redistribution of sound energy, a listener stationed south of the source will be able to hear the sound at relatively great distances. It is left for the reader to determine what the acoustical result would be for a listener located to the north of the sound source.

3–5. *Diffraction*

When a wave disturbance of any nature is incident on an aperture in a screen, some energy will pass through the opening. **If the**

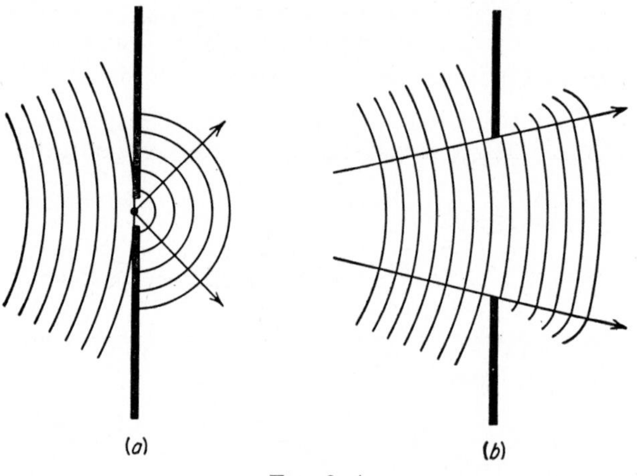

Fig. 3–4

aperture is small compared to the wavelength, the wave disturbance will spread out into the region behind the screen. In such a case the opening in the screen acts, in certain respects, as a source of sound energy for the region behind the screen. Such a situation is shown diagrammatically in Fig. 3–4*a*. If the opening is relatively large, as in (*b*), little spreading will occur. A similar phenomenon may also appear at the edge of a screen. This spreading out of a wave disturbance behind a screen, or behind its edge, is known as **diffraction.**

In the field of musical acoustics we are dealing with waves varying in length from about 3 cm to something like 10 m. If the opening from which the sound energy issues is of the order of 30 cm in diameter, sounds having a frequency of, say, 8000 cps (λ = 4.3 cm) will spread out (be diffracted) very little into the region behind the non-transmitting surface, while the sounds having a frequency of the order of 100 cycles (λ = 343 cm) would be propagated in all directions behind the screen.

A practical example of this type of diffraction is to be found in the performance of the loudspeakers used in so-called public-address systems, particularly in those of the horn type. The voice or musical sound issuing from a loudspeaker is composite in nature, i.e., it consists of sounds of various frequencies. A listener located on the principal axis of the horn hears the sound of all frequencies in something like their proper relative intensity, but an auditor sitting near the side of the auditorium probably hears only the lower frequency tones. As a result of diffraction, the "highs" are more or less eliminated, with the result that the voice or music sounds unnatural.

The second type of diffraction (the bending of the sound wave about the edge of an obstruction) is exemplified when one sits behind a tall or large person in a theater. The direct sound waves are more or less completely absorbed by the body of the person in front. The sound energy that does reach the person sitting in the sound shadow reaches him as a result of diffraction around the outline of the individual in front.

3–6. *Reflection of Sound*

When a sound wave is incident upon any given surface, either or both of two things may happen—reflection and absorption. It is the reflected energy with which we are concerned at present. When a sound wave undergoes reflection from any surface, the angle which the reflected ray makes with the normal to the point of incidence is equal to the angle which the incident rays make with the same reference line. This is illustrated in Fig. 3–5.

In (*a*) of this illustration, *NP* is the normal at the point of incidence P. The angle R is equal to the angle I. This relation holds under all circumstances, and is of great importance, as will be seen shortly. There are two types of reflection—**diffuse** and **regular.** If the reflecting surface is rough, that is, if the surface irregularities

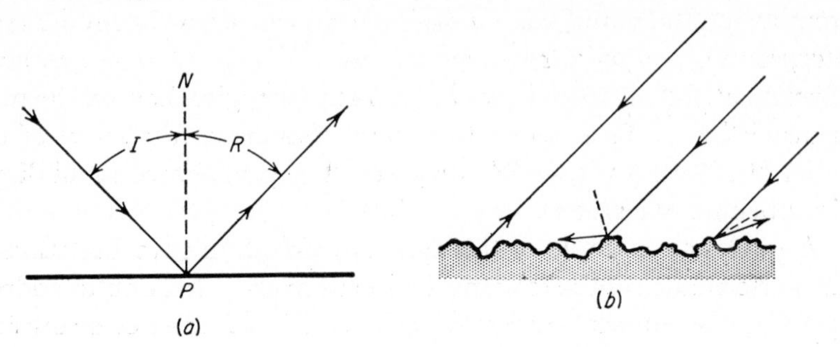

Fig. 3–5

are comparable in size with the wavelength involved (Fig. 3–5*b*), diffuse reflection will take place; if, however, the surface is relatively smooth, regular reflection will occur. The reflection of a spherical wave from a plane surface is shown by the sound photograph reproduced in Fig. 3–6.

Fig. 3–6

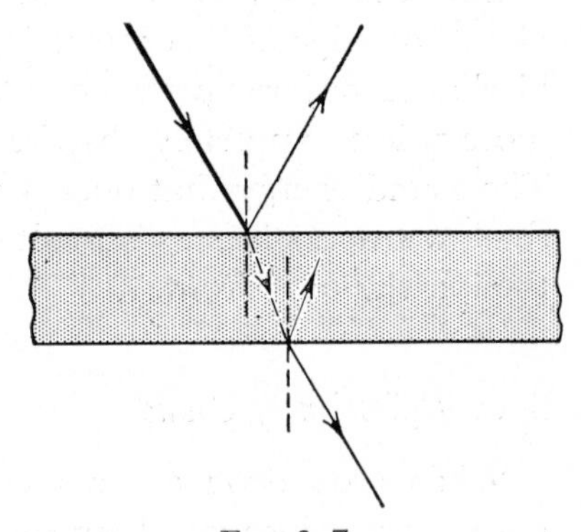

Fig. 3–7

The percentage of the energy content of the incident wave which is reflected depends upon the nature of the reflecting surface. When a train of sound waves is incident on a solid, rigid body having a relatively smooth surface, reflection will be maximum. Any energy which enters the receiving body will be partly absorbed and partly transmitted. The general situation is sketched in Fig. 3–7. The broken line within the medium is intended to indicate that repeated reflection takes place from the layers or particles constituting the body of the medium. The width of the lines roughly represents the relative amount of energy which is reflected, absorbed, and transmitted. Certain materials will absorb a high percentage (as much as 90 per cent in some cases) of the energy content of the incident wave.

For the present, however, we are concerned with that part of the incident energy that undergoes reflection. There are various striking examples of the reflection of sound, as for instance in the Mormon Tabernacle in Salt Lake City. In this instance the building is more

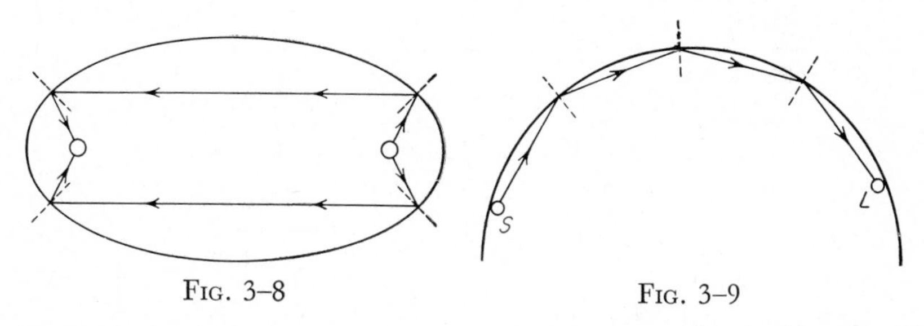

FIG. 3–8

FIG. 3–9

or less elliptical in cross section, with the reading desk at or near one of the foci of the ellipse. An auditor stationed at the far end of this large building and near the other focus can hear the sound made by the dropping of a pin on the reading desk. The two curved end surfaces act as elliptical sound mirrors, as roughly sketched in Fig. 3–8. The basic laws of reflection hold in this and other curved surfaces as they do in the case of plane surfaces.

Multiple reflections from a single curved surface account for the acoustical behavior of so-called whispering galleries. In such cases the unusual loudness with which a whisper may be heard is not confined to any one position but may be noticed at various points near the wall of a room having a curved wall or ceiling. The principle involved in such cases is illustrated by the sketch shown as Fig. 3–9. Successive reflections of the sound waves originating at, say, *S* may reach the listener at *L*, and thus serve to augment the sound reaching this point by direct radiation. The sound thus appears to "creep" along the wall.

Certain church interiors have a more or less domelike ceiling, with the result that there often exist in such buildings one or more focal points where sound is abnormally loud. Such concave reflecting surfaces give rise to a nonuniform energy distribution and hence result in poor room acoustics.

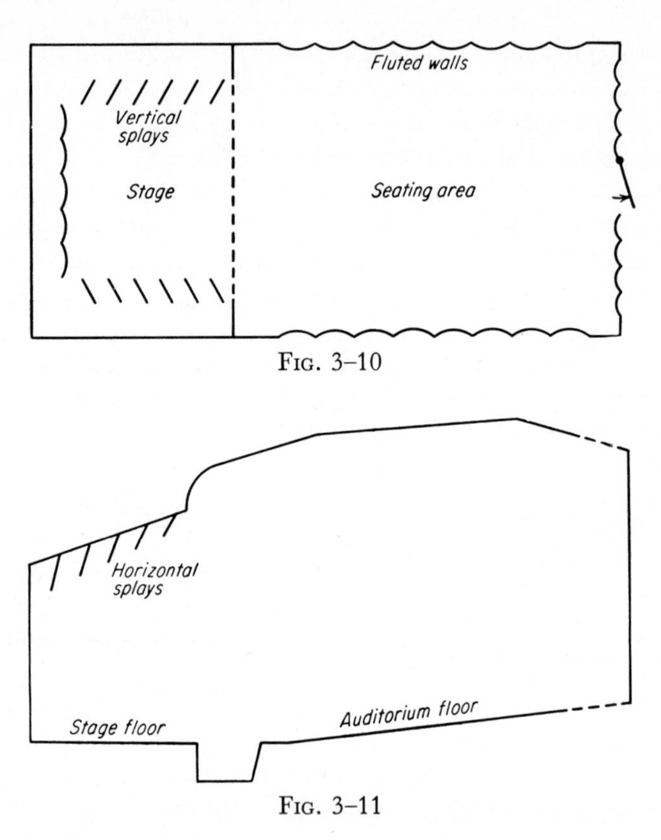

Fig. 3–10

Fig. 3–11

By the same token it is evident that convex surfaces may be utilized for the purpose of diffusing the available sound energy and thus securing a more uniform sound distribution than would otherwise occur. The walls of a theater or concrete hall can, for instance, be made up of a series of convexly curved sections, as sketched in Fig. 3–10. To bring about the reflection of sound toward particular seating areas, the side walls and the ceiling of the stage or band shell can be splayed, as indicated in Figs. 3–10 and 3–11.

When utilizing specifically designed surfaces for purposes such as those above indicated it becomes necessary to recognize the fact that the reflective process is, to some extent at least, a selective one. A surface of limited dimensions does not reflect all wavelengths with the same effectiveness. As we know, most sounds are complex in character, that is, they are made up of a group of waves having various frequencies. Certain of these wave components will be reflected from a surface having given dimensions, while others of a different wavelength will be reflected only to a slight degree. It may be said, in general, that maximum reflection occurs when the dimensions of the reflecting surface are large compared with the wavelength of the incident sound. This means that, in the case of a music auditorium where use is made of special reflecting surfaces, both the curved and the flat sections must have a surface width at least as great as that of the longest wavelength involved in the music being rendered. Further, the radius of curvature of the curved wall sections should be so chosen that uniform diffusion in a given room will result.

Occasionally use is made of the laws of reflection in connection with public speaking. A curved reflecting surface is sometimes placed behind the position occupied by a minister as he stands in the pulpit. When properly designed and positioned such a reflecting structure will redirect a considerable amount of sound energy toward the auditors and thus serve to augment the direct radiation. However, in more recent years electroacoustical equipment is rapidly replacing such reflector installations. Public-address systems will be discussed in a later chapter.

There is one other special case of reflection that is worthy of note here. There are conditions under which sound waves are caused to travel within a rigid tube. In dealing with such cases one must keep in mind a certain fundamental principle involved in connection with wave propagation. This basic principle is that, in general, reflection of waves takes place if and when a wave train encounters a boundary where there is a change in the density of the medium. In dealing with waves propagated through tubes two possible situations may obtain. In one case the far end of the tube may be closed by a rigid wall. As a wave train arrives at such a closed end a pressure change will occur; the pressure in a compression area will increase and that in a rarefaction region will decrease. These changes in pressure give rise to reflected waves; a compression is thus reflected as a compression and a rarefaction as a rarefaction.

However, in the case of an *open* tube the situation is somewhat different. If the end of the tube toward which the waves are traveling is open, the air within the tube is free to expand outward. Under these conditions when a compression arrives at the open end, the pressure becomes less. Likewise, when a rarefaction reaches the open end, it becomes less of a rarefaction than it would have been if the tube extended farther. This change in pressure, then, gives rise to reflected waves; but in this case the compression is reflected as a rarefaction, and the rarefaction as a compression. In other words, at the open end a change of phase has occurred. This is a very important fact; and these two cases should be kept in mind when we later take up the subject of acoustical resonance.

3–7. *Absorption*

As we shall see later, it is often desirable to reduce reflection to a minimum. To accomplish this a surface, such as a wall or ceiling, may be covered with a material that is of such nature and construction that it absorbs a large percentage of the incident sound energy. Carpets, heavy curtains, and various special materials may be used as absorbing media. The effectiveness of a given material in absorbing sound energy is referred to as its **absorption coefficient.** By this is meant the percentage of energy absorbed as compared to the total incident energy. For example, if a material absorbs 60 per cent of the incident energy and reflects 40 per cent, the absorption coefficient will be 0.6. Each square foot of the material is the equivalent of 0.6 sq ft of a perfectly absorbing surface. In practice it is the custom to consider an open window as a perfect absorber, and to take 1 sq ft of such an opening as the basis of acoustical absorption calculations. This unit is called the sabin, after Professor Sabine. It signifies the absorption due to 1 sq ft of a perfectly absorbing surface. On this basis various materials are rated as to their absorbing effectiveness. On page 33 is a table giving the absorption factor number of representative materials.[1] Since the

[1] For an extended list of coefficients of sound-absorbing materials, the reader is

absorption coefficient varies with pitch, the values given are based on a frequency of 500. In general, the absorption factor is lower for the lower frequencies.

Material	Description	Absorption coefficient $f = 500$ cps
Carpet	Unlined	0.20
Carpet	Heavy, with lining	0.37
Curtains	Heavy, draped	0.50
Acousti-Celotex, C-2	Perforated cellulose fiber tile, ⅝″ thick	0.69
Acousti-Celotex, M-1	Perforated mineral tile, ⅝″ thick, incombustible	0.50
Fiberglas Acoustical Tile, TXW	Textured surface, ¾″ thick	0.69
Fiberglas Acoustical Tile, PRW	Perforated surface, ¾″ thick	0.75
Fibertone	Perforated cellulose fiber tile, ¾″ thick	0.73
Permacoustic	Fishered mineral tile, ¾″ thick, incombustible	0.75
Plaster	½″ thick	0.04–0.06
Wood	Varnished	0.03
Wood paneling		0.06
Wood veneer	On 2″ × 3″ wood studs, 16″ c/c	0.12
Brick wall	Unpainted	0.03
Brick wall	Painted	0.17
Concrete	Unpainted	0.16
Marble		0.01
Glass		0.027
Audience	Units (sabins) per person seated	4.0±

An examination of the above table reveals the fact that such wall materials as ordinary plaster, wood, and brick absorb very little of the incident sound energy; nearly 100 per cent is reflected. However, in the case of the various special acoustical materials, something like 75 per cent of the energy is absorbed.[1] There are

referred to Knudsen, "Architectural Acoustics," pp. 198, 240ff. A somewhat later list is to be found in Knudsen and Harris, "Acoustical Designing in Architecture," Appendix 1.

[1] For any given material, the absorption secured will depend upon the thickness of the material and also upon the method of its mounting.

many such materials commercially on the market, the surface of some of which may be decorated without destroying their absorbing property; and a number of them are fireproof. By the use of one or more of the various sound absorbing materials it is possible to control the amount of acoustical energy which is reflected from the walls and ceiling of an auditorium. The desirability of such control is discussed in Sec. 17–1.

3–8. *Echoes*

The reflection of sound waves gives rise to the familiar phenomenon of echoes. If a reflecting surface of considerable size is located at a distance, say, of something like 100 ft, and if the plane of the reflecting surface is at right angles to the direction of propagation of waves produced by a sharp sound impulse, one will hear not only the original sound but also, after a time interval of approximately $2/10$ sec, a delayed replica of the primary sound, this second sound being the *echo* of the original disturbance. If the hearing is acute it is possible to distinguish a distinct echo even when the reflecting surface is only about 50 ft away from the source of the sound. In the open, the reflecting surface may be the outside wall of a building or the side of a sheer cliff. In such places as the Grand Canyon, marked and long delayed echoes are caused by reflection from distant rocky cliffs. Inside, a wall or ceiling may function as such a sound mirror. In a long narrow room, successive reflections may take place as the sound waves are reflected back and forth between the parallel walls, thus producing an acoustical "flutter" that diminishes in loudness with each reflection. Often such echoes take on the characteristics of a musical sound having a definite pitch. Such a phenomenon is the result of successive reflections from closely adjacent and parallel reflecting surfaces, such as the risers of the concrete seats in an athletic stadium or the pickets of a fence. In such cases the time interval between successive reflections is so small that the resultant sound takes on the character of a musical note.

Several practical applications of the echo phenomenon strikingly illustrate the laws of reflection. One such instance has to do with a method whereby bodies submerged in water or at depths in the earth may be located. In recent years geophysical measurements have, to an increasing extent, come to replace the older and more expensive hit-or-miss methods of drilling in order to locate oil-bearing and other earth strata. One such method consists in causing one or more explosions near to or in the surface of the earth. The resulting sound waves as they travel outward from the source through the earth's surface will undergo reflections if and when they encounter a boundary between two media whose transmission characteristics are different. Diffraction may also occur. The speed of the direct and reflected sound waves will depend upon the nature of the substrata. In such an exploratory operation the *modus operandi* consists in locating several sound pickup devices (special microphones) at various points some distance from the sound center and simultaneously recording by electrical means the time at which the reflected and refracted sound waves reach the various pickup stations. Since the speed of sound through various earth formations is fairly well known, it thus becomes possible to determine the distance, direction, and character of the various earth strata in a given area.

Radar equipment which serves to warn aviators of the near-presence of other planes or of land masses also utilizes the phenomenon of echo formation. Waves of very high frequency (ultrasonics) are radiated in short pulses by a special radiating unit. When striking a solid object these waves give rise to echoes. The reflected waves are detected by electronic means, thus making it possible to determine the transit time. When this is known, the distance of the reflecting object can be quickly computed, and its presence thus be made known.

QUESTIONS

1. In what respect does a sound wave differ from a water wave?

2. Is it possible for a wave to travel at a high speed and at the same time for the vibrating particles to have relatively small average speed?

3. What is the relative speed of sound in a solid, such as wood, and in a gas? Give the reason for your answer.

4. Assuming that the speed of sound is 1100 ft/sec, what will be the wavelength of a sound whose

frequency is, for example, 523 cps?

5. If the speed of sound is 1100 ft/sec at a given temperature, what will be its speed if the temperature rises 25°F?

6. The wavelength of a certain sound in air is 5 ft. What will be the wavelength of the same sound when transmitted through water? Steel?

7. Of what practical significance is the fact that sound wave may be refracted?

8. A boy claps his hands in front of a public library building, and hears a momentary musical sound reflected from the steps leading to the building. What is the frequency of the sound if the steps are 1 ft wide, and the temperature is 68°F?

9. If an echo is heard 0.2 sec after a sound is made, how far away is the reflecting surface?

4 *Interference*

4–1. *Theory of Wave Interference*

When two or more sets of wave trains simultaneously traverse the same medium, they may give rise to certain effects known as interference. In order to clearly understand this important phenomenon it will be well to consider somewhat further the nature of wave

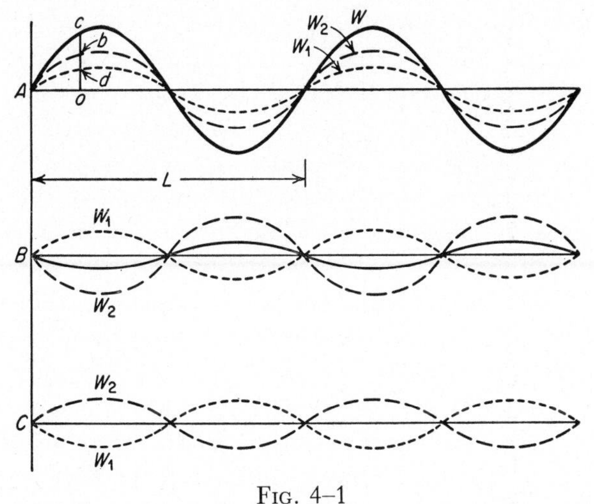

FIG. 4–1

motion, in general, and more particularly the result which follows when two or more periodic motions are combined.

Suppose we have two coexisting sinusoidal waves W_1 and W_2 as sketched in Fig. 4–1. The two waves are of equal length, and they are **in phase,** but they are of unequal amplitude. In order to determine the form of the resultant wave we add, algebraically, the ordinates (perpendiculars) of the constituent waves W_1 and W_2 in A; that is, we add *od* to *ob* and get the corresponding point *c* on the resultant wave W. If this is done for a number of points along the time axis, the complete resultant waveform may be drawn. It will be noted that the resultant amplitude is greater than the amplitude of either of the components, and that the resultant wave is sinusoidal and in phase with both W_1 and W_2.

In the case illustrated in B of the same figure, W_1 and W_2 are in **opposite phase.** Proceeding as above, we find that the resultant wave has an amplitude which is **less than that of the smaller of the two component waves.** Here we have a case of partial interference.

In C we have the case of two wave systems of **equal** amplitude and period but of opposite phase. Carrying out the process of algebraically adding the ordinates, we see that the result is zero. In other words, one wave system completely nullifies the other. This is an instance of **complete** interference.

Above we have been speaking of wave motion in general. Sound waves, it will be remembered, consist of alternate regions of compression and rarefaction, and not of "troughs" and "crests" as pictured above. However, as we have seen, longitudinal waves, such as those of sound, may be graphically represented in the same manner. In the case of sound waves the areas above the time axis represent areas of compression, while the areas below correspond to conditions of rarefaction. On this basis it will be evident that if and when two sound waves of equal amplitude and frequency but of opposite phase are simultaneously traversing the same medium, **the result will be silence.** We shall see later that circumstances may be such that two wave trains will arrive at certain points out of phase by a half period while at other points in the medium they may be only slightly out of phase. In such cases complete nullification will obtain at certain points and only partial or no interference at others. But in any case where interference occurs **energy is not destroyed; it is redistributed.**

4–2. *Examples of Interference*

The destructive interference of sound waves may be experimentally demonstrated in a comparatively simple manner. If one excites an ordinary tuning fork and slowly turns it as it is held near the ear, four positions will be found at which the sound will be practically inaudible. In general, as the fork vibrates, the prongs move outward and inward simultaneously. When they move outward, for instance, their motion gives rise to regions of compression outside the prongs and an area of rarefaction between the prongs. These regions of compression and rarefaction come together along four lines,[1] as indicated in Fig. 4–2, and these periodic disturbances are thus out of phase by a half period; hence they nullify one another, and silence

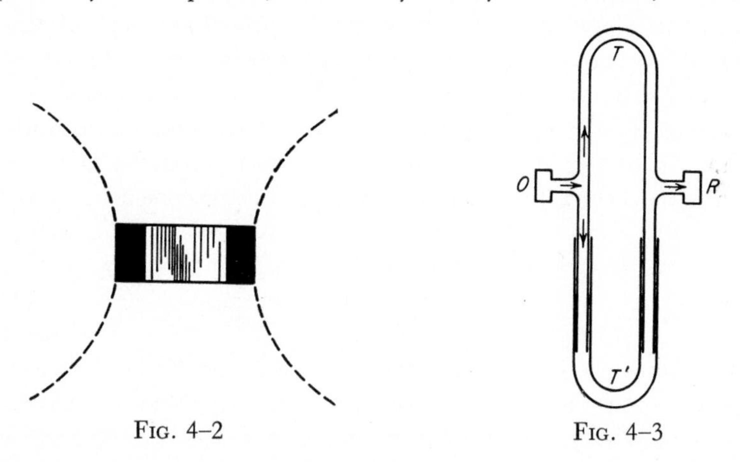

FIG. 4–2

FIG. 4–3

is the result. As further proof that the observed effect is due to interference, one may place a pasteboard tube carefully about one of the prongs and note the fact that the fork may be heard with equal loudness in all positions.

An additional example of the principle of interference is to be had by making use of a divided tube as sketched in Fig. 4–3. T and T' together constitute a divided sound path for any sound produced at O. T' slips over the remainder of the tube much as in the case of the slide trombone. Any convenient source of high-pitched (why?) sound may be used at O, and some type of microphone as a receiving agent at R. A high-impedance telephone receiver driven by a beat-frequency oscillator answers well for the source O, and the energy picked up by the microphone at R may be amplified and made audible by means of a loudspeaker. A stethoscope may be used instead of the microphone and speaker.

[1] It has been shown by Weber that these lines form equilateral hyperbolas.

The sound waves divide as shown by the arrows, and, if the paths T and T' are of equal length, will reach R in phase, and sound will be heard. If, however, T' is made longer than T, by means of the adjustable feature, the waves traversing T' will arrive at R out of phase with those coming via T. The amount of phase difference will depend upon the difference in path. If this difference in path amounts to **one-half a wavelength,** the wave trains will reach R out of phase by one-half a period, and complete interference will obtain. Under these circumstances no sound will be heard. This experiment not only illustrates the phenomenon of interference in a striking manner, but it also brings out another important fact, viz., that **phase difference may be brought about by causing two wave trains from a common source to traverse paths of unequal length.** This aspect of interference has a bearing on certain problems connected with the acoustics of rooms, as discussed in Chap. 17.

4–3. *Beats*

In dealing with the theory of interference in Sec. 4–1, it was assumed that the periods of the two sources of sound were equal; that is, the number of excursions per second described by each sonorous body was the same. We come now to a case of interference in which the operating conditions are somewhat different than in the examples already considered. Actually we seldom deal with two musical sounds of strictly equal pitch.[1] It will therefore be of more than passing interest to examine the case where we have two sources of sound of nearly equal pitch. What will be the result when two sounds of unequal pitch simultaneously traverse a given medium? The answer to this question carries implications which are of great importance to all musicians.

If we take the two resonant forks and sound them simultaneously, we hear a single sound which is materially louder than that heard when one of the forks is sounded alone. If now we change slightly the pitch of one of these forks by attaching to one of its prongs a bit of wax and again excite both forks, an interesting and highly

[1] For the present "pitch" may be thought of as synonymous with "frequency." Later a distinction will be made between these two terms.

significant phenomenon presents itself. It will be noted that the sound, which before was continuous, gradually dying down as the forks lost energy, now periodically rises and falls in intensity. These pulsations in intensity, which we call **beats,** are caused by alternate additive and subtractive interference. Periodically the two wave systems arrive at any given point **out of phase** by a half period with the result that almost complete destructive interference obtains at this point. The totality of maximum interference lasts for only a very small fraction of a second, leading to the illusion that the sound is continuous and only varies in intensity. As a matter of fact, the sound will not reach zero intensity at any given point unless the amplitudes of the two sounds are strictly equal, and unless the interfering wave trains are exactly opposite in phase.

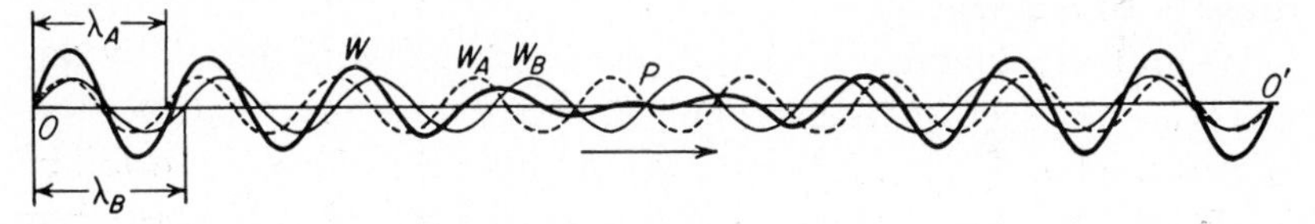

FIG. 4–4

A better understanding of the theory of beats may be had from an examination of the diagram sketched in Fig. 4–4, which represents the conditions at some particular instant.

Let us assume that the unloaded fork A above referred to gives rise to a sinusoidal waveform as represented by curve W_A in the diagram, and that the loaded fork B gives rise to a wave of equal amplitude (for the sake of simplicity) whose graph would be W_B. Since the frequency of the second fork B has been made less than the first A, its wavelength will be correspondingly greater. It will therefore be obvious that a greater and greater phase difference will appear as one proceeds from O toward the right. Between O and P there are four and one-half wavelengths of W_A and only four of W_B. If now we add, algebraically, the amplitudes of the two wave trains, we get the resultant wave W. Near the beginning of our observations (in the region of O) the amplitude of the resultant is greater than that of either of the components; this is due to the fact that W_A and W_B are nearly in phase. However, in the region of P the amplitudes add up to practically zero, the two waves being out of phase by a half period. Since distances above the zero line represent magnitudes of compression and those below the X axis rarefactions, it will be seen that in the region of P a condition of condensation meets one of rarefaction. The intensity is therefore zero. As one proceeds toward the right, the difference in wavelength results in a gradual diminution in phase difference until at O' the two wave trains are again completely in phase, and augmented intensity results. Since both of these wave trains are in progress toward the right, this interference phenomenon will appear to a listener stationed at some fixed point as a series of periodic pulsations in intensity. In other words, the auditor is conscious of what we have already designated as beats. **Beats may therefore be defined as the periodic variations in the intensity of sound at a point due to the coexistence of two wave trains having slightly different frequencies.**

When the frequency of fork A differs from that of fork B by one vibration per second, there is one alternation, or one beat, per second. If the two sounds differ in pitch by five vibrations, there will be five beats per second. In general, it may be said that **the number of beats per second is equal to the difference in the frequencies of the two primary sounds.** When the beat frequency is greater than about twenty per second it is not possible to distinguish the individual beats but the resultant sound may, under certain circumstances, take on a "rough" character—a phenomenon which we shall examine later. It should be noted that beats are variations in wave amplitude and are not themselves waves. As we are here using the term beat, there is no such thing as an **objective** beat tone.

In the preceding discussion we have assumed that the sounds involved in the production of beats were sinusoidal in character, that is, the tones were simple. But it has already been noted that most musical tones are quite complex, consisting of a group of sounds having various frequencies. If two or more complex tones are sounded simultaneously, beats may be caused by the out-of-phase components of two such tones. Later we shall consider the effect of this type of interference on the character of musical sounds.

4–4. *Use of Beats*

The existence of beats makes it possible to adjust two sonorous bodies to a common frequency; that is, to determine when two notes are in unison. In order to accomplish this, the sonorous body under test is adjusted until it produces no beats when sounded simultaneously with a source of standard pitch. By such means it is possible to distinguish between the pitch of two notes that differ from each other in frequency by a fraction of 1 vps. This principle is made use of in connection with piano and organ tuning. A detailed discussion of this tuning procedure will be included in the section dealing with temperament in Chap. 9.

The phenomenon of beat formation is sometimes utilized in connection with the production of organ tones. Certain organ stops, such as the voix céleste, consist of two ranks of pipes one of which is tuned to a slightly different pitch than the other. When a given key is depressed two pipes will speak, and, because they differ slightly in pitch, slow beats will occur. The result is to introduce into the tone a slow undulating, wavelike effect. This effect is considered, by musicians, to give to the resulting note a pleasing tonal color.

To a certain extent a beat effect obtains when a particular section of an orchestra plays alone, as for instance the string choir. In such a case, probably no two violins are in perfect unison. Thus the combined effect of fifteen or twenty violins is to produce a multitude of slow beats. These beats give to the music rendered by a string ensemble, for instance, a slightly undulating effect which contributes a certain satisfying quality to the music that would otherwise be lacking. The sound of a single violin could be amplified to equal in loudness the combined effect of many violins, but the musical effect produced by a single instrument would be far less acceptable than that produced by a group of players. The existence of beats is undoubtedly a prominent factor in producing this interesting and significant effect.

4–5. *Standing Waves*

We conclude our discussion of interference with a reference to a certain interesting and useful type of interference phenomenon. **Standing** or **"stationary" waves** in a gaseous medium are formed by the combining of two sets of sound waves of equal amplitude and wavelength which are moving in opposite directions. They are called standing waves because the waveform does not progress.

This type of acoustical disturbance may exist when a **reflected** or secondary wave train is superimposed upon a direct or primary radiation, under proper conditions. Reference to Fig. 4–5 will make clear how this may come about.[1] Let us assume that S is a source of sound and W a reflecting surface. Let A represent a group

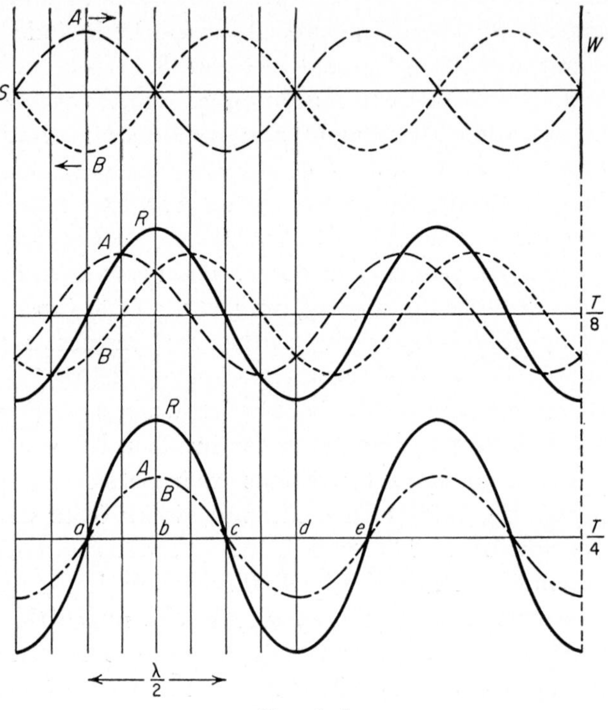

FIG. 4–5

of sound waves undergoing propagation toward the right, as indicated by the arrow, and B a corresponding wave train **after reflection.** If we superimpose these two wave trains at the instant shown in the sketch, the resultant displacement of the medium would be zero. One-eighth of a period later A will have moved toward the right a distance equal to one-eighth of a wavelength, and B will

[1] It should be remembered that, in the case of sound waves, the actual displacement of the medium is parallel to the direction of propagation. In representing the case graphically, however, it is more convenient to sketch the case as if the displacement were taking place transversely.

have progressed an equal distance toward the left. They will thus have shifted their relative position by a quarter wavelength, as shown in the second group of curves in the diagram. If now we add the amplitudes, algebraically, we get the resultant waveform *R*. The situation one-eighth of a cycle later ($t = T/4$) is shown in the lowest group of curves. Here the wave *A* has advanced toward *W* a distance equivalent to one-fourth of a wavelength, and *B* has advanced toward the left an equal distance. The waves will accordingly have moved with respect to one another a distance equal to **one-half a wavelength.** Sketching in the resultant wave by the usual procedure, we see that while its amplitude has increased, **its position has not changed.** If we were to take the time and space to diagram the remainder of the cycle, it would be found that the resultant waveform would always remain fixed in space, and thus we have what has already been referred to as a standing, or stationary, wave. Such a wave has two important characteristics: (1) certain particles of the medium such as those at *a*, *c*, *e*, etc., a half wavelength apart, **are always at rest;** and (2) at certain other points such as *b*, *d*, etc., the particles of the medium, also a half wavelength apart, **execute a periodic motion** whose amplitude is twice that of each of the individual wave trains. The points *a*, *c*, *e*, etc., are referred to as **nodes,** and the points *b*, *d*, etc., as **antinodes.**

The existence of standing waves may be readily demonstrated in a number of ways. One simple and effective method is to make a whistle out of a wooden mouthpiece and a glass tube closed at the opposite end, as roughly sketched in Fig. 4–6. The glass tube forming the body of the whistle may conveniently be about 15 inches long and one inch in diameter. A small amount of cork dust introduced into the tube will serve to indicate the position of the nodes and antinodes when the whistle is sounded. On blowing such a whistle, we see a striking example of standing waves; the cork dust will show definite regions of marked agitation corresponding to areas of compression. When the sound ceases, the dust will be found to lie in sharply defined ridges corresponding to the nodes with spaces between at the antinodal points. The distance between

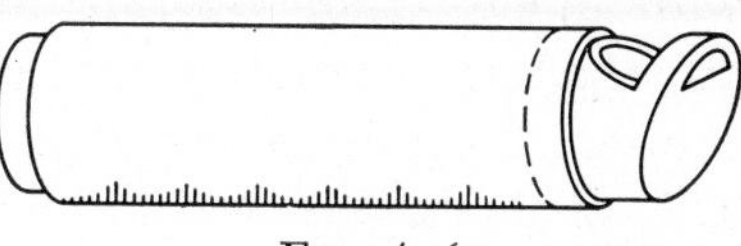

FIG. 4–6

two adjacent ridges correponds to one-half a wavelength. From such a measurement one might compute the frequency of the sound by means of the relation, given in Sec. 3–1,

$$f = \frac{s}{\lambda}$$

where f represents the frequency, s the speed of sound, and λ the wavelength.

There are several variants of the simple experiment just described, that may be utilized as a means of finding the speed of sound in various media. A convenient method, utilizing standing waves, recently adopted by some investigators makes use of an electrically generated tone and an electrical pickup device, somewhat as outlined in connection with the interference studies discussed in Sec.

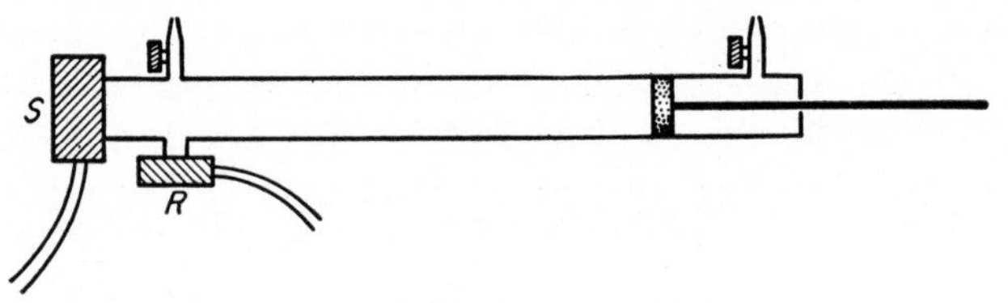

FIG. 4–7

4–2. Such an organization for measuring the speed of sound in air and other gases is diagrammatically sketched in Fig. 4–7. The apparatus consists of a metal tube, one end of which is provided with an electrically operated tone generator *S*, and the other with an adjustable reflector for altering the length of the vibrating column. Near the generator end is attached an electrical pickup *R*. This pickup serves as a means for determining the location of the nodal and antinodal points as the position of the reflector is changed. The distance between positions of the reflector when the pickup *R* indicates maxima (antinodes) and minima (nodes) enables the experimenter to determine the wavelength of the sound emitted by the generator. If the frequency of the source is known, the speed can at once be computed. The presence of the side tubes makes it possible to fill the tube with various gases. The sound generator is usually supplied with alternating current from a microphone hummer or a beat-frequency oscillator, and the output of the sound detector is amplified and passed through a suitable indicating meter. This method yields more accurate results than the older methods.

In the foregoing discussion we have confined our treatment to a consideration of standing waves in a gaseous medium. It should however be noted that standing-wave phenomena may also exist in solid media. The periodic mechanical vibrations of a sound generator, such as a violin string or a bell, may develop standing waves within the structure itself. When we come to the study of the vibrations of strings and other sonorous bodies we shall consider the subject of standing waves in such cases.

QUESTIONS

1. Assuming a standard source of sound whose frequency is 440 cps, what would be the frequency of another sonorous body when 4 beats per second are heard?

2. Suggest how you would determine whether the frequency of the second body, in question 1, was greater or less than that of the standard.

3. How is the phenomenon of beats used in musical work?

4. Suggest an experimental procedure whereby one might make use of the standing-wave effect to determine the speed of sound in a gas, and in a solid.

5 *Hearing*

5–1. *Structure of the Ear*

Having discussed the nature of sound and the process by which sound waves are propagated through a conducting medium, we may now address ourselves to the question of how such waves are detected and caused to give rise to a physiological sensation.

Nature has provided us with a sound-detecting mechanism—the human ear. As a preliminary step in the study of the process of hearing it will be in order to glance at the anatomical structure of the organ of hearing. The general structure of the ear is sketched in Fig. 5–1. The hearing mechanism is made up of three parts: the outer, the inner, and the middle ear. The external auditory canal (the meatus, 2) is closed at its inner end by the "eardrum" (membrana tympani, 3), consisting of a thin layer of fibrous tissue covered with skin externally and with mucous membrane internally. The tympanic membrane is somewhat elliptical in shape, its major axis being horizontal.

The middle ear is filled with air at atmospheric pressure and houses a chain of three small bones referred to as the auditory ossicles, 5, 6, and 7. Because of their shape, these important structural components are named: the hammer (mallus), the anvil (incus), and the stirrup (stapes). The ossicles serve to connect, mechanically, the drum with the inner ear structure. The "handle" of the hammer is attached eccentrically to the inner surface of the tympanic membrane, and is so articulated with that member that the drum is pulled slightly inward, thus causing the drum to be slightly concave when viewed from the outside. The base of the stirrup is attached to the middle-ear assembly by means of an elastic membrane that closes an opening (the oval window) in the wall of the middle ear. The footplate of the stirrup is nearly as large as the oval window, and is positioned a little off center. In the normal ear the ossicles are not rigidly connected with one another. The equality of air pressure between the outer and the middle ear is maintained by means of the Eustachian tube, which opens into the upper part of the throat in the postnasal region.

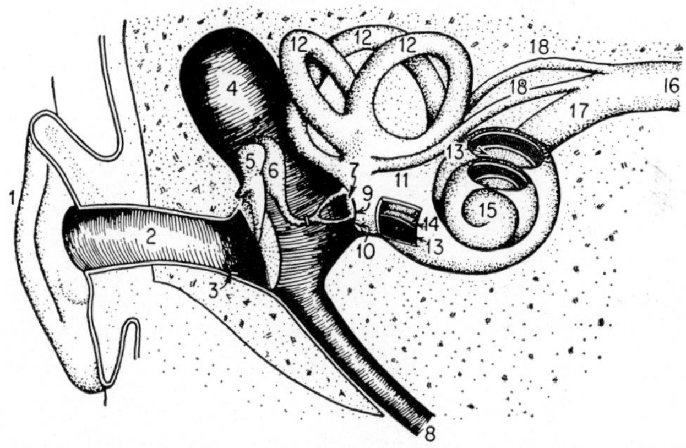

FIG. 5–1. Diagrammatic view of the human ear. 1, pinna (external ear); 2, auditory meatus; 3, tympanic membrane (drum); 4, middle ear; 5, malleus; 6, incus; 7, stapes; 8, Eustachian tube; 9, fenestra ovalis; 10, fenestra rotunda; 11, vestibule; 12, semicircular canals; 13, scala tympani; 14, scala vestibuli; 15, cochlea; 16, auditory nerve; 17, cochlear branch; 18, vestibular branch. (Courtesy of Prof. R. A. Waggener, Carleton College.)

The inner ear constitutes, in some respects, the most remarkable structure of the human body. It consists of three major parts: the semicircular canals (12), the vestibule (11), and the cochlea (15). The vestibule, as indicated by its name, serves as an entrance to the semicircular canals and the cochlea. The three semicircular canals apparently serve to give us a sense of balance, but they do not appear to function in connection with the hearing process; hence they will not be considered further.

The cochlea (15), as the diagram indicates, is a snail-shaped structure consisting of a tapering tubular cavity of about two and a half turns, and measuring about 3.3 mm in diameter at its base; its total length is about 32 mm. A diagrammatic sketch of the cochlea is shown in Fig. 5–2. It is seen to consist of three galleries extending throughout its length, except for a small connecting passage, the helicotrema, near the apex of the structure. Structurally these chambers are formed by a bony partition (lamina spiralis) extending about two-thirds of the way across the main chamber, and to which are attached two membranous structures which in turn are attached to the outer wall. The horizontal member is called the basilar membrane, and the other the membrane of Reissner. These two membranes divide the cochlea into three

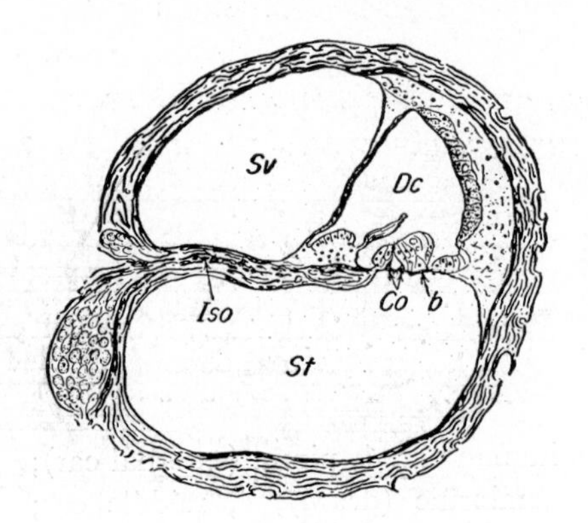

FIG. 5–2. Diagrammatic section through one coil of the cochlea. *St*, scala tympani; *Sv*, scala vestibuli; *Dc*, ductus cochlearis; *Iso*, lamina spiralis; *b*, membrana basilaris; *Co*, rods of Corti.

chambers: the lower one, the scala tympani; the larger of the upper regions, the scala vestibuli; the smaller one, the ductus cochlearis. Since the membrane of Reissner is very thin and flexible, from the hydrodynamic point of view, the two upper chambers may be considered as a single chamber. Each of these chambers, and the vestibule, are filled with a liquid, the endolymph. The scala vestibuli is physically separated from the inner ear by the oval window (9) in Fig. 5–1. The scala tympani terminates in the round window (10) which consists of a small area covered by flexible membranous tissue. A diagrammatic sketch of the cochlea, uncoiled, is shown in Fig. 5–3.

The basilar membrane and its associated components constitute the most important part of the hearing assembly. Its structure is highly complex and we shall refer only to certain of its details. This membrane is 0.21 mm wide at the smaller end, and consists of several thousand fiberlike elements arranged transversely side by side. On the upper surface of the basilar membrane is to be found the organ of Corti. This consists of several specialized elements, chief of which are the *rods of Corti* and the associated nerve terminals

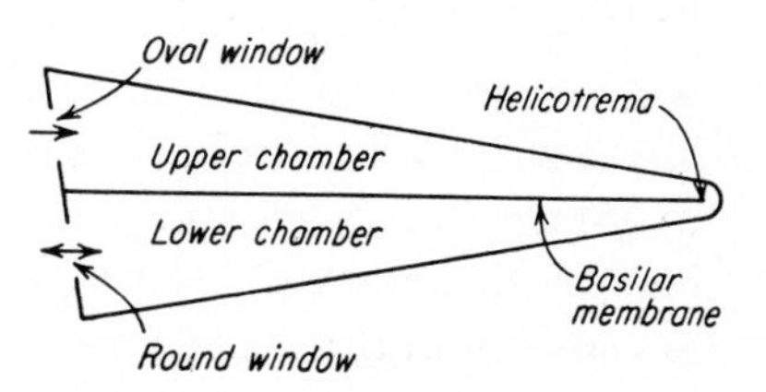

FIG. 5–3. Diagrammatic sketch of cochlear structure.

in the form of small hair cells. (See Fig. 5–4 for details.) Altogether there are some 23,500 of these so-called rods. A hair cell is to be found at the outer end of each rod, and from each such cell twelve or fifteen hair cilia project into the liquid filling the cochlea. Lying over these cilia, and touching them, is a soft and loose structure

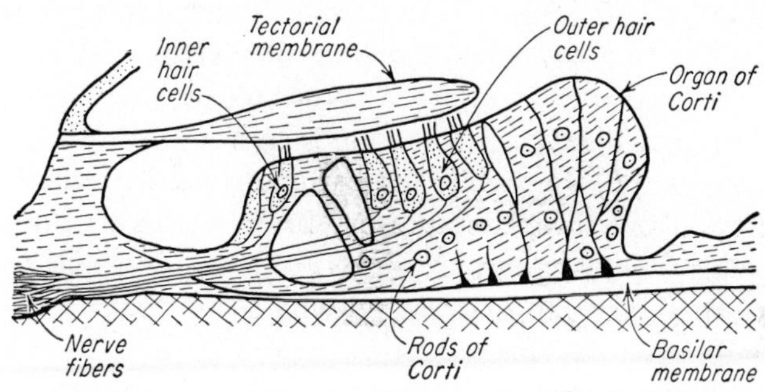

FIG. 5–4. Enlarged sketch of organ of Corti and associated parts.

known as the tectorial membrane. The auditory nerve, a branch of the eighth cranial nerve, enters the central axis of the cochlea as indicated in Fig. 5–2. This nerve structure consists of a bundle of several thousand nerve fibers forming a tiny "cable" that is about a millimeter in diameter. Each nerve fiber appears to be connected to five hair cells. This nerve system passes to the brain structure through the temporal bone.[1]

[1] The reader is referred to Stevens and Davis, "Hearing," p. 268, for a more detailed discussion of the anatomy of the inner ear, or to Fletcher, "Speech and Hearing," p. 111, for a general description of the organ of hearing.

5–2. *The Hearing Process*

Having considered the anatomical structure of the ear, we are now in a position to give consideration to the process whereby the detecting device that we call the ear makes it possible for sound waves (a mechanical motion) to give rise to the sensation of hearing. Up to a certain point in the process the evidence of what happens is reasonably definite and well understood, but, as we shall see, beyond that stage the exact *modus operandi* of the process is, to some extent, still in the controversial stage. In order to discuss the subject of hearing in detail it would be necessary to have recourse to a study of neurophysiology, biophysics, and related subjects. It must therefore suffice to review only briefly the latest thought in this field.

To begin with, sound waves entering the hearing structure through the auditory canal (meatus) impinge on the tympanic membrane (drum) causing it to vibrate. This membrane is acoustically sensitive to a remarkable degree. It is estimated that a movement of the eardrum through a distance of the order of a millionth of a millimeter will give rise to a sound sensation. It should be noted, in this connection, that while the eardrums are extremely sensitive to changes in pressure, the auditory brain center will not be activated unless the change in pressure is repeated in rapid succession. Later we shall come to understand why this is so.

Any vibration of the eardrum gives rise to a corresponding motion of the ossicles, and hence a transfer of energy to the membrane covering the oval window. In this energy-transfer process an interesting and important mechanical relation presents itself. Since the ossicles act as a lever system, and since the area of the plate of the stapes is only about one-twentieth of that of the eardrum, the pressure exerted per unit area on the liquid in the middle ear is from thirty to sixty times greater than that exerted by sound waves on a corresponding area of the eardrum. It is thus to be seen that the drum and the mechanism of the middle ear together function as a form of mechanical transformer that serves to transfer effectively the energy content of the air-borne sound waves to the liquid medium of the middle ear.[1]

[1] To those who are acquainted with the basic laws of mechanics it will be evident

Up to this point in our discussion we are on fairly firm ground, but the last stage of the hearing process is not so definite. A substantial amount of research work has been done in an effort to determine just how the pressure waves communicated to the oval window are transformed into nerve impulses. Owing to the small size of the inner ear structure, and to its location, the experimental work involved in the study of the inner ear, in the case of a live subject, is extremely difficult. Perhaps the most thorough study of the hearing process made in recent years was carried out by Professors Stevens and Davis, of Harvard University. Briefly stated, the results of their studies, and the research of other investigators, lead to the following tentative conclusions. The basilar membrane and its associated anatomical components constitute what might be called "the seat of hearing." It is in this rather complex assembly that the conversion of physical energy into nerve impulses takes place. The variable pressure exerted by the stapes sets up a pressure wave in the fluid of the inner ear. Since this fluid is essentially incompressible, and since the cochlear fluid is enclosed in a rigid capsule, nature has provided a pressure release device, viz., the fenestra rotunda (round window). The elastic membrane covering this window serves as such a pressure compensating component, thus making it possible for pressure waves to exist in the endolymph. The pressure waves developed in the endolymph give rise to a corresponding motion of the basilar membrane and its associated parts, particularly the cilia protruding from the hair cells. As the tectorial membrane is in contact with these cilia, it is evident that this relative motion will cause the cilia to undergo a bending motion. This vibratory motion of the cilia serves to excite the hair cells, thus instituting nerve impulses. It is found that these nerve impulses are generated only during the outward movement of the eardrum. The nerve impulses thus developed progress along the nerve fibers, eventually reaching the appropriate brain center, thus giving rise to the sensation of hearing. While the elements of the basilar membrane are to some extent progressively tuned, the vibratile elements of the inner ear are, in general, mechanically damped, which fact precludes sharp tuning. Pitch perception appears to depend upon

that this transformer action serves to match the impedance of the outer ear to that of the middle region.

the response of one or more zones of the basilar membrane and its associated parts. In general it may be said that the wave-responsive elements near the broad end of the cochlea respond to the higher frequencies, while those near the apex pick up the lower frequencies. The frequency response of the normal ear extends from about 20 cps to approximately 20,000, a range of about 10 octaves. The response limits differ somewhat with different individuals, the average upper limit being about 18,000. In old age the response to the higher frequencies tends to fall off somewhat.

The cochlear response to frequency is complex, being approximately linear for faint sounds but nonlinear for loud sounds. In a later section we shall in fact see that the aural response bears a logarithmic relation to the magnitude of the exciting energy. This is fortunate because the ear is called upon to function through an extremely wide dynamic range.

In concluding our discussion of the ear it may be said that students of aural acoustics have not yet arrived at a complete working explanation of the physiological process involved in the neural activity of the auditory system. All living cells exhibit certain electrical properties, and this phenomenon has recently been utilized by research workers in their study of the hearing process. The hair cells of the cochlea exhibit certain electrical characteristics and also show certain reactions when stimulated electrically, and it seems quite probable that valuable information will be secured as a result of these studies.[1]

Deafness may be due to one or more causes. Speaking broadly, there are two principal types of deafness: transmission-deafness and nerve-deafness. The former may result from any cause that impairs transmission of the energy content of sound waves to the inner ear. This may involve physiological changes in the tympanic membrane, or the ossicles, or the presence of material that tends to in-

[1] H. Davis, The Electrical Phenomena of the Cochlea and the Auditory Nerve, *J. Acoust. Soc. Am.*, April, 1935. This paper also carries a comprehensive bibliography of research done in this field up to that date. Two more recent papers are: G. V. Békésy, DC Potentials and Energy Balance of the Cochlear Partition, *J. Acoust. Soc. Am.*, vol. 23, no. 5, pp. 576–582, September, 1951; and G. V. Békésy, DC Resting Potentials Inside the Cochlear Partition, *J. Acoust. Soc. Am.*, vol. 24, no. 1, pp. 72–76, January, 1952.

hibit their normal movement. The other type of auditory defect is known as nerve-deafness and results from degeneration of the auditory nerve proper or of the hair cells of the organ of Corti. This form of deafness cannot be corrected. In the case of an impairment of the hearing function caused by defective transmission through the middle ear assembly, it is possible to bypass these components by making use of sound conduction through the bones of the head, and thus to secure some improvement in the hearing process.

We have dealt with the ear somewhat at length because of its importance as the acoustical detector with which nature has provided us whereby the energy of sound is converted into physical sensation. As we proceed with our study of musical acoustics, it is desirable to have an acquaintance with both the capabilities and limitations of the hearing mechanism and its functioning.

5–3. *Intensity of Sound*

As pointed out in Sec. 3–1, the energy radiated by a sonorous body is conveyed to the reviewing agent by means of a wave motion in some material medium. The effect due to a traveling wave involves not only the magnitude of the energy content of the wave at any given point, but also the time rate at which that energy is available at that point. These two aspects of the situation are embodied in the concept referred to as the intensity of the wave motion. As applied to sound, **we may define the intensity of the sound wave as the average time rate at which acoustic energy passes through a unit area normal to the direction of wave propagation.** Such a rate of energy flow might be expressed in ergs per second per square centimeter. Recalling the discussion appearing in Sec. 1–10, we see that we are here dealing with the concept of power, and that such a time rate of energy flux may therefore be expressed in watts per square centimeter. The foregoing observations lead to the statement that sound intensity is the acoustic power per square centimeter of the sound wavefront. Since the power represented by a train of sound waves is, in general, quite small, a smaller unit than the watt is commonly used when dealing with acoustical problems. Such a unit is called a **microwatt,** which is one-millionth of a watt. If one were to speak softly, for instance, the sound intensity would be something like one-tenth of a microwatt/sq cm. The average public speaker develops sound energy at the rate of 25 to 50 milliwatts. For purposes of comparison, it may be noted that the average electric light bulb is rated at 40 watts; this is equivalent to 40 million microwatts. It would require something like 1500 bass voices singing fortissimo to supply enough energy to light a 40-watt bulb.

It may be shown that the intensity of a given sound, at any point, is indicated by the relation

$$I = 2\pi^2 a^2 n^2 ds$$

where a is the amplitude, n the frequency, d the density of the medium, and s the speed of sound. From the above it is evident that for a given frequency and a particular medium the intensity of a sound varies directly as the square of the amplitude. Now the amplitude depends upon three factors:

1. The amplitude of vibration of the sonorous body
2. The superficial area of the sounding body
3. The distance from the source

In the case of the last mentioned factor, it may be said that, in a free sound field (absence of reflecting surfaces), the intensity varies inversely as the square of the distance from the source. If, for instance, the distance from the source is doubled, the intensity will drop to one-fourth of its former value. The second factor mentioned above is particularly significant. The statement may be readily verified by a simple experiment. If one excites a tuning fork the sound is quite faint—the intensity is low. If however the shank of the vibrating fork is held in contact with the top of a table, the intensity of the sound will be considerably increased. The increase is due to the fact that the table top, which was thrown into vibration by the fork, has a much greater superficial area than the fork. But in this connection a word of caution must be sounded. **The total available energy was not increased by placing the fork on the table.** Careful observation of the experiment will disclose the fact that the fork when in contact with the table will stop vibrating much sooner than when held in the hand. This is because the fork communicated some of its energy to the table; **the rate at which the fork delivered energy was increased,** hence the increase in intensity.

In practice, sound intensity, i.e., the time rate of energy flow (power), is measured in terms of pressure changes in the medium due to the compressional wave. It has been established that in such cases the flow of energy per second per square centimeter (power/cm^2) is given by the expression

$$I = \frac{p^2}{ds}$$

where p is the effective pressure change (in dynes/cm^2), d the density of the medium (in gm/cm^3), and s the speed of the compressional wave (in cm/sec). In air at 20°C, it is found that the above expression for intensity (I) reduces to the form

$$I = \frac{p^2}{415}$$

Thus we have a relation by which one may compute sound intensity in microwatts per square centimeter. The sound pressure (p) can be determined by experimental means. Since, as indicated above, the intensity of sound waves decreases inversely as the square of the distance from the source, and since the intensity varies directly as the square of the pressure, it follows that **sound pressure varies inversely with the distance.**

In the foregoing discussion we have been dealing with absolute intensity magnitudes. In the next section we shall consider relative intensity values.

5–4. *Intensity Level—The Decibel Scale*

The acoustician is usually interested in comparing the intensities of two sounds rather than in the absolute value of either. In doing this it has come to be the practice to deal with the ratio of the two intensities involved. As previously mentioned, the human ear is a remarkable physical organ. It responds to an extremely wide range of intensity. In fact, the ear will respond to a sound whose intensity is 10 billion times that required to produce a just audible sound. Because of this wide range of sensitivity, it has been found convenient to make use of a logarithmic[1] scale in comparing sound intensity. Happily, the logarithmic-scale idea fits into another important aspect of the intensity picture. There is a general relationship known as the Weber-Fechner law to the effect that the response of any sense organ is proportional to the logarithm of the magnitude of the stimulus. As applied to the sense of hearing this would mean that if one were comparing two sounds, one which had an intensity of 100 units with another whose intensity was 10 units, the aural response in the first case would be twice that due to the less intense sound. This is because the logarithm of 100 is 2 while the logarithm of 10 is 1.

In comparing intensities it becomes necessary to assume a standard of reference. Having done this, we may set up an expression for what is called the **intensity level** of a sound. Using the logarithmic relation above referred to, one may write a defining equation that takes the form

$$N = 10 \log_{10} (I/I_0)$$

where N is the intensity level in decibels (db)* and I_0 the assumed reference intensity. In the above relation both the intensity (I) under comparison and the reference intensity (I_0) are expressed in watts or microwatts per square centimeter.

Because of the relations previously discussed, it is also possible to set up a corresponding relation in which the comparison is made on the basis of pressure instead of intensity. This relation is

$$N = 20 \log_{10} (P/P_0)$$

where N has the same significance as above, P the sound pressure being considered, and P_0 the reference sound pressure. In this case the sound pressures are expressed in dynes per square centimeters.

[1] For the benefit of those who are not familiar with logarithms, it may be said that a logarithm is the power (exponent) to which a given number must be raised in order to give another number. For example, since $10^2 = 100$, and $10^3 = 1000$, 2 is the logarithm of 100, 3 is the logarithm of 1000.

* The unit of intensity level is known as the bel, named after Alexander Graham Bell, the inventor of the telephone. Since the bel is a rather large unit, in practice a unit whose value is one-tenth of the bel is commonly used; it is called the decibel (abbreviated db).

When air is the medium, it has been found that for a sound whose frequency is 1000 cps, the minimal power per unit area, in a plane progressive wave, that will evoke a perceptible response in the ear is 10^{-10} microwatt. This is therefore taken as the reference intensity (I_0). The corresponding sound pressure is 0.0002 dyne/sq cm, and this value accordingly serves as a reference sound pressure (P_0). When plane progressive waves are involved, both of the above equations will yield the same numerical result.

In considering the matter of intensity levels it is to be noted that a change in sound intensity of one decibel is approximately the smallest change in energy content that can ordinarily be recognized by the human ear. This corresponds to a change in acoustic power of approximately 26 per cent. In other words, the intensity of a given sound must be increased by one decibel, or 26 per cent, before the ear can detect any change in the "strength" of the sound. When the intensity of sound reaches a value of 120 db the listener begins to experience pain, and if the ear is subjected to a sound of such intensity for a considerable period of time, damage to the organ of hearing may result. The very intense sounds occurring in connection with some industrial operations may thus become a health hazard. The following table gives the intensity levels of various sources of sound as given by different investigators.[1]

NOISE LEVELS COMMONLY ENCOUNTERED

Source of sound	RMS pressure in dynes/cm²	Intensity level in db
Sound becomes painful	200	120
Hammer blows on steel, 2 ft	80	114
Airplane engine	50	110
Diesel engine 2-cyl, 200 hp, 5 ft	30	105
Riveter, 35 ft	15	97
Heavy traffic	6.4	90
Noisy factory	3.6	85
Average factory	1.6	78
Ordinary conversation, 3 ft	0.36	65
Average office	0.062	50

[1] See a paper by the author entitled Industrial Noise Control, appearing in *Gas J.*, p. 143, October, 1953.

Among the significant comparisons that might be made is the fact that at a point in an auditorium or concert hall where one can, for instance, just hear the violin section when playing softly (at an intensity level of about 1 db) the sound from the full orchestra playing fortissimo may have an intensity of 85 db. This means that the acoustic power involved in the two cases would be of the order of 20 million to one.

In musical and architectural acoustics it is often necessary to measure the intensity level of a given sound. Equipment is available for the purpose of carrying out such measurements. Below (Fig. 5–5) is shown one commercial form of such an instrument,

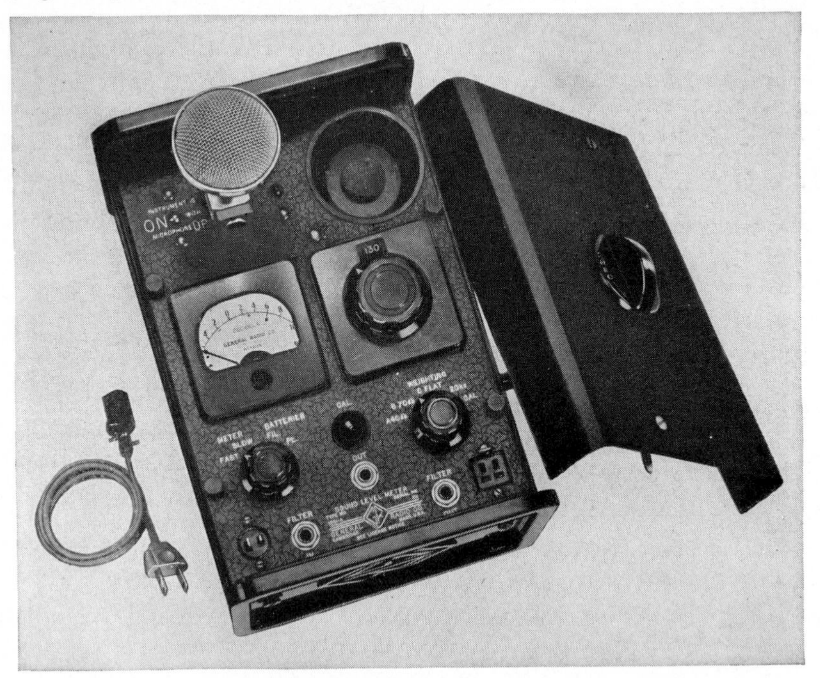

FIG. 5–5. Sound-level meter. (General Radio Co.)

known as a "sound-level meter." The essential components of such an assembly consist of a suitable microphone, an amplifier, a calibrated attenuator, and an indicating meter reading in decibels. The pressure of the sound waves under study actuates the microphone, thus converting the acoustical energy into electrical currents which in turn serve to operate the db meter.

5–5. *Loudness*

It has been said that most arguments would terminate at once if the disputants were to define the terms involved. Such an observation is particularly apropos when one examines the literature bearing on the subject of loudness. In the previous section it was pointed out that the term intensity indicated the magnitude of a sound, expressed in terms of pressure or energy, and as measured by instruments. Loudness has to do with the **subjective** aspects of the hearing process, that is, with **magnitude of a sensation as judged by a given individual.** Here we are dealing with a phenomenon in the realm of psychophysics. For this reason it is extremely difficult to arrive at anything resembling a quantitative approach to the problem involved. However, as a beginning, we might set up a rough definition of loudness by saying that the term is used to designate the magnitude of the sensation experienced by an auditor when a sound wave impinges on the eardrum. Almost any listener having normal hearing can roughly classify the magnitude of auditory sensations as "very loud," "loud," or "moderately loud," "soft," or "very soft." In musical terminology such designations are indicated by ff, f, mf, p, or pp, respectively. But these terms are not and cannot be precise, because an aural response depends upon the individual, and upon other factors such as pitch. Indeed the two ears of a given individual often exhibit unequal responses. However, in dealing with acoustical problems it is desirable, if possible, to establish some method of expressing loudness in terms that will have at least some quantitative significance. If a fixed relation existed between aural response (sensation) and the magnitude of the stimulus (intensity) the case would be comparatively simple, but unfortunately no simple relation exists. For instance, at the upper and lower intensity limits there is an increasing departure from the logarithmic relation previously referred to. What then shall be our approach to the problem of establishing some sort of scale of loudness?

Plainly, any scale of loudness that might be set up should agree with our common experience in estimating sensation magnitudes. Such a scale should show, for instance, that when the units are doubled, the sensation will be doubled, and when the scale reading is trebled the sensation will be trebled, etc. A number of investigators, including Fletcher and Munson, Geiger and Firestone, and Churcher, have attacked the problem of setting up a usable loudness scale. As a result of experimental research, such a scale has been established. In Fig. 5–6 may be seen a graphic representation of the relation existing between the intensity of a sound and the subjective loudness. Here the intensity is expressed in decibels above the hearing threshold and the loudness in loudness units. Some writers use the terms sones and millisones as loudness units (1 sone = 1000 millisones = 1000 loudness units). The curve shows that at the

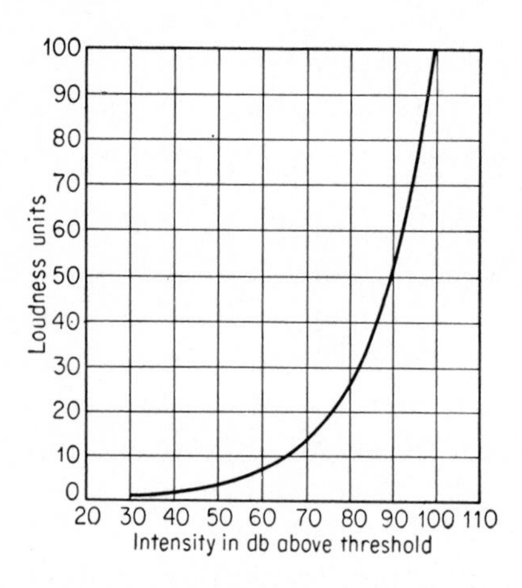

Fig. 5–6. Graph showing relation between intensity of sound and subjective loudness. (Adapted from a chart in "Theory of Hearing" by E. G. Wever, 1949, by permission of John Wiley & Sons, Inc.)

lower intensities a given change in intensity produces a much smaller change in loudness than one in the region of the higher levels of intensity.

The effect that frequency has on loudness in the case of pure tones is indicated by the curves shown in Fig. 5–7. An examination of the graph discloses the fact that frequency does not appreciably affect loudness when the tone falls between about 800 and 2000 cycles, but that at frequencies beyond this range the loudness decreases as the frequency increases. The curves also show that at the lower frequencies, such as 100 cycles, for instance, the loudness falls off rapidly as the frequency decreases.

There is another way by which the loudness relation may be represented, and by which the sensitivity of the ear at various frequencies may be made apparent. This can be done by setting up a series of equal-loudness contours, as shown in Fig. 5–8. In setting up these curves a pure tone of 1000 cycles, and having an intensity of 10^{-10} microwatt/sq cm, is taken as a basis of comparison. The loudness level of this reference tone is taken to be the same numerically as its intensity level. On this basis the loudness level of any pure tone is determined by listening to the sound in question and also to the standard 1000-cycle tone, alternately, and adjusting the

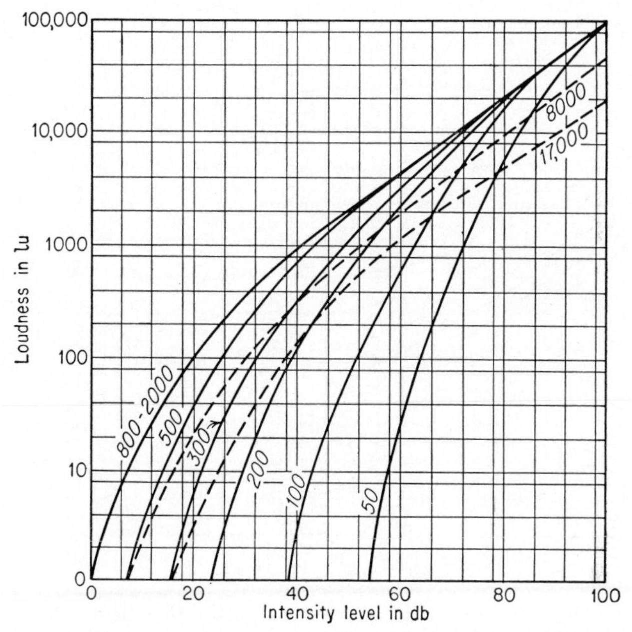

FIG. 5–7. Graph chart showing the relation between the intensity level and the subjective loudness of pure tones of various frequencies. (Reproduced by permission of Dr. Harvey Fletcher and *The Journal of the Acoustical Society of America.*)

strength of the standard tone until the two sounds are sensed to be of equal loudness. The loudness level of the sound under test is then given by the intensity level of the standard or reference tone. The numbers appearing on the curves indicate loudness levels in decibels. Since the decibel is also used for the unit of intensity, less confusion will result if some other name be utilized as a loudness-level designation. The term phon is accordingly used for this purpose. From what has been said above, it will be evident that the loudness level expressed in phons will be numerically equal to the loudness level expressed in decibels. For example, the contour line marked 40 in Fig. 5–8 might be labeled 40 phons. The curve marked zero indicates a reference loudness of 0 db; it outlines the threshold of audibility. It is to be noted that, by definition, in the case of the 1000-cycle tone the loudness levels agree, numerically, with intensity levels. However, it will also be seen that at other frequencies

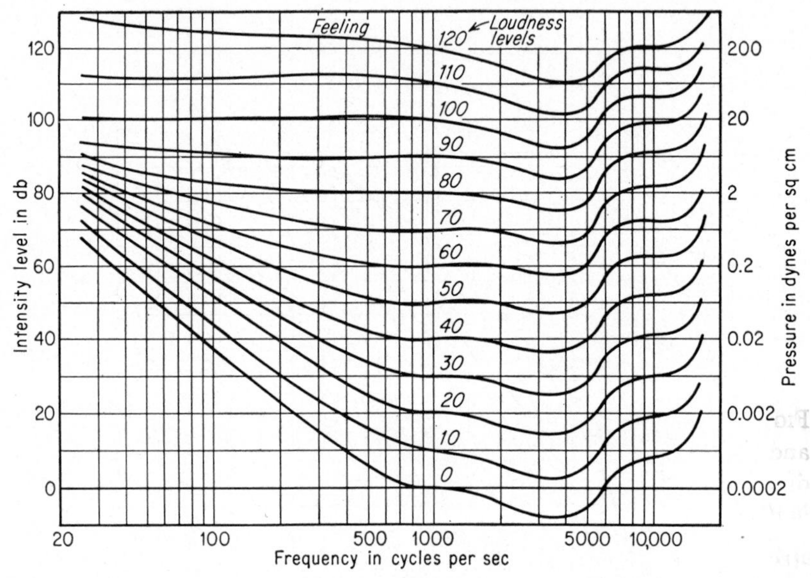

Fig. 5–8. Loudness-level curves. Adopted by the American Standards Association, 1936. The reference level is 10^{-10} microwatts/sq cm. Sound pressure given at right.

the loudness values do not equal the intensity values. For instance, it will be seen that at a frequency of 200 cycles a sound must have an intensity of at least 25 db before it will be audible, while a note whose frequency is, say, 400 would be audible if the intensity were only 10 db. The upper contour indicates that when the loudness level is above 120 we become conscious of the wave disturbance through the sense of feeling rather than by hearing. Various useful comparisons can be made by the use of such a loudness contour chart. Now it is right at this point that confusion is likely to enter into the discussion. The contour curves do not indicate subjective loudness. They simply serve as a convenient means of comparing the ear response at different frequencies.

In concluding our discussion of the subject of loudness, it should be pointed out that two complex tones having the same fundamental frequency and the same intensity may differ widely in subjective loudness. A careful study of the data presented in Figs. 5–7 and 5–8 will disclose why this is so. We shall return to this aspect of the loudness question when we take up the matter of tone quality.

5–6. *Sensitivity of the Ear*

As previously indicated, the average normal ear can hear sounds between the frequency limits of about 20 and 20,000 cps. In a few cases the upper limit may be as high as 25,000. But the upper frequency limits drop with advancing years; in the 40- to 50-year bracket it commonly decreases to something like 15,000, and in old age it may fall below 10,000. For the moment, however, we are not concerned so much with frequency limitations as we are with the intensity-loudness relation. An inspection of the lowest contour curve of Fig. 5–8 discloses the fact that the ear is most sensitive at a frequency of about 3000 cps. At that frequency a subjective sound reaction will result if the effective pressure of the sound wave is less than 0.0002 dyne/sq cm, while at the lower and the higher frequencies the sensitivity falls off rapidly. The minimal pressure above indicated corresponds to a sound intensity of less than 10^{-4} micromicrowatt/sq cm. It is estimated that the amplitude of vibration of the eardrum under such a condition would be of the order of 10^{-8} mm. It is thus seen that the human ear exhibits a truly remarkable sensitivity.

If the intensity of sound is increased, the subjective loudness increases and at length reaches a level at which the reaction is one of feeling rather than of auditory response. This is indicated by the higher contour line of Fig. 5–8, and it will be noted that this level, known as the "threshold of feeling," is less dependent on frequency than is the minimal threshold. However, this upper limit of audition varies from one person to another.

If one connects the ends of the contour lines above mentioned, as shown in Fig. 5–9, there is found an acoustical region known as the auditory sensation area. Each point in this area represents a particular auditory sensation when the ear is receiving sound waves having the intensity level and the frequency value indicated by the corresponding coordinates. Sounds that are of a complex nature cannot be represented by a single point on the auditory diagram.

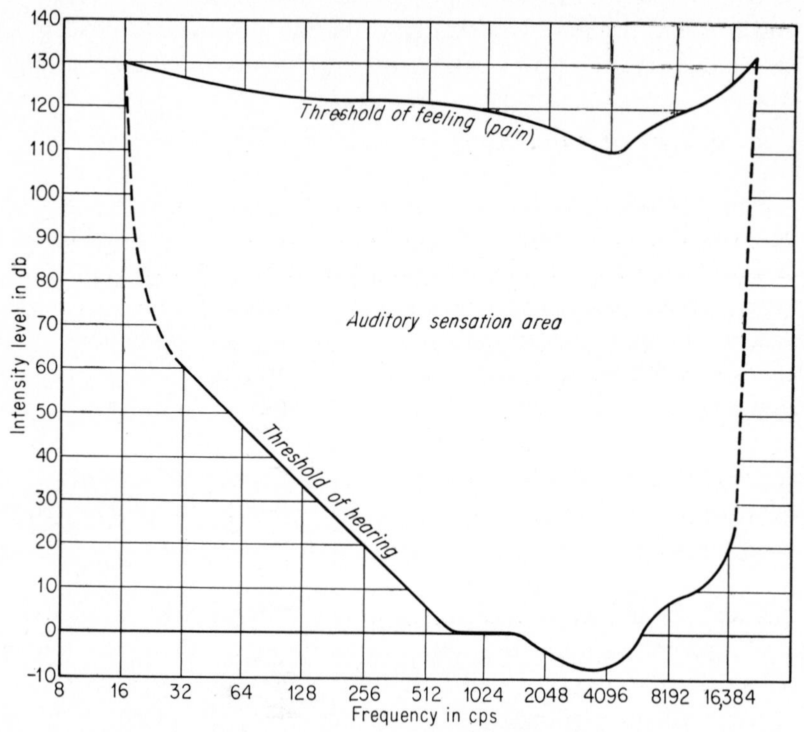

FIG. 5–9. Chart showing average response of the normal ear to sounds of different frequency. (After Steinberg, Montgomery, and Gardner.)

Sounds that lie outside of this region are not heard. For persons who are partially deaf, the threshold of hearing contour lies higher in the diagram, but in such cases the upper curve remains about the same.[1] Some people who would not be considered deaf in the ordinary sense of the term, nevertheless have defective hearing. For instance, a person is occasionally found whose ears are insensitive

[1] Typical audiograms in cases of deafness are to be found in H. Fletcher, "Speech and Hearing," p. 200, and should be consulted.

to some particular band of frequencies. To such individuals the quality of musical sounds would seem different than to a person having normal hearing, as we shall see later.

There is another aspect of the sensitivity question that is significant both in the field of music and in connection with problems involved in architectural acoustics. Reference is here made to the idea of differential intensity-level sensitivity. The size and shape of a person's auditory sensation area, as discussed above, will yield significant data concerning his hearing limitations, but another aspect of the auditory picture, in any given case, has to do with minimal change in intensity that can be detected by the listener. For instance if the frequency of a given sound is 200 cps and its intensity is 60 db, what is the smallest change in intensity that the observer can detect? It is found, for example, that if an individual can become conscious of a perceptible loudness change when the intensity of a pure tone changes, for example, from 60 db to 60.8 db, or from 59.2 db to 60 db, that person's differential intensity sensitivity would be 0.8 db. Tests show that the value of the differential sensitivity depends upon both the intensity of the incident waves and the frequency involved. From the above it is obvious that one's aural reaction to musical and other sounds will, to some extent, be determined by one's differential sensitivity.

Equipment is available by the use of which a person's auditory diagram may be determined experimentally. Such a unit is known as an **audiometer,** and consists essentially of an audio oscillator, an attenuator, an amplifier, and a telephone receiver. The oscillator produces an alternating electrical current of small magnitude and controlled audio frequency. These minute currents are strengthened by the amplifier and fed into the telephone receiver. The amplified electric currents cause the receiver to emit a pure tone to which the subject listens in the usual way. The attenuator, which is in effect a resistor, is connected between the oscillator and the amplifier and serves to control the strength of the current fed to the telephone receiver, and hence the intensity of the sound issuing from this component. The attenuator is calibrated in decibels. In practice certain frequencies are commonly used in such tests, viz., 128, 256, 512, 1024, 2048, 4096, and 8192 cps. The reference level of intensity is that given by the lowest contour line in Fig. 5–8. If the person whose hearing is being tested has normal hearing, the attenuator will show a series of reference readings at each of the test frequencies. If, however, the auditory response of the person being tested is below normal, the attenuator will have to be so adjusted that a more intense sound will issue from the receiver—to a level that is audible to the subject. The difference between the attenuator

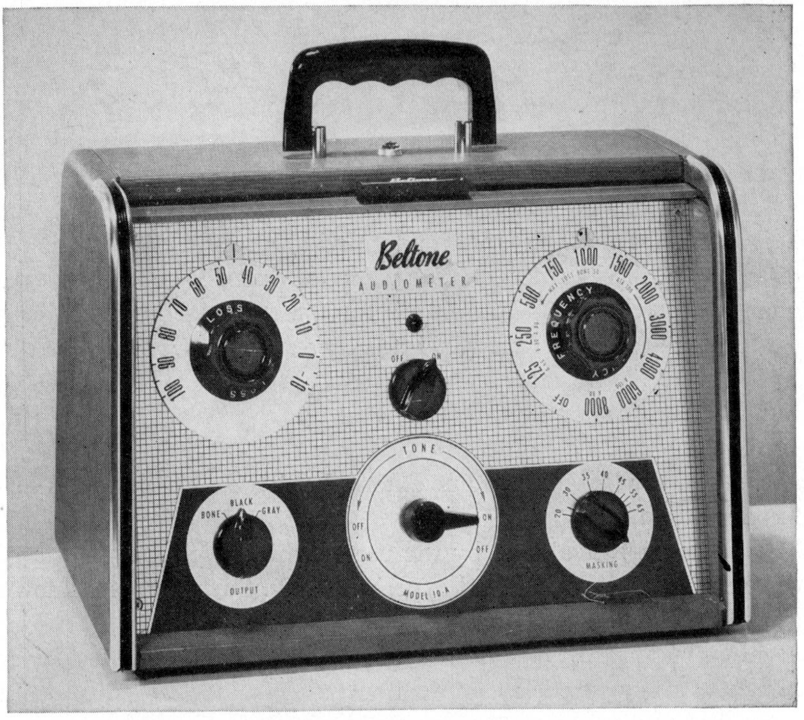

Fig. 5–10. Audiometer. (Beltone Hearing Aid Co.)

reference reading and the threshold reading as shown by test will then be the hearing loss at that particular frequency of the person being tested. When readings at the various test frequencies have been made, a graph can be constructed and superimposed on the normal hearing diagram, such as that shown in Fig. 5–9. The data thus secured will yield much important information concerning one's auditory response to musical and other sounds. Fig. 5–10 shows a representative audiometer assembly, and in Fig. 5–11 is displayed typical test-response curves.

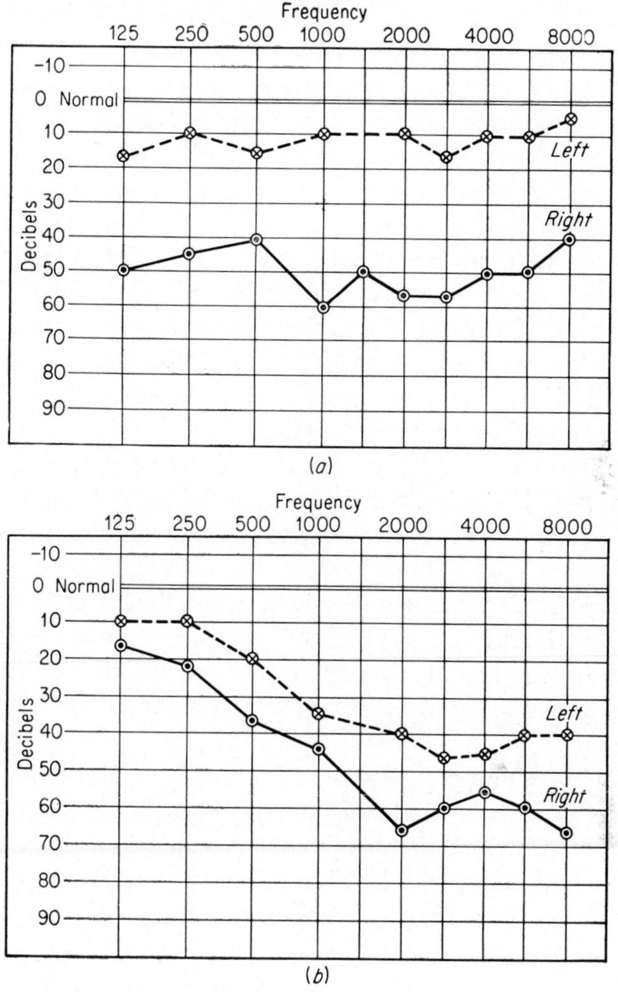

FIG. 5–11. Hearing response curves: (*a*) left ear normal; hearing in right ear seriously impaired; (*b*) case of congenital deafness in both ears. (Courtesy of Hazel S. Whitson, audiometrist.)

5–7. *Masking*

There is a certain phenomenon that has an important bearing on auditory results, particularly in connection with one's judgment of musical quality, to be discussed later.

It is a more or less common experience that the existence of one sound may affect the ability of the ear to hear another sound. If, for example, a note whose frequency is 200 cps is sounded simultaneously with another note whose frequency is, say, 2500 cps, and if the 200-cycle note is gradually increased in intensity while the higher note is held at a constant intensity level, there comes a time when the higher note can scarcely be heard. Such an effect is referred to as **masking.** So far as the perception of the higher pitched sound is concerned, the presence of the masking sound has the effect of raising the threshold of hearing. It is also true that a high-frequency tone may serve to mask a tone of relatively low frequency, particularly if the frequencies of the two sounds are near together. It has been found that the effectiveness of the masking depends upon both the frequency and the intensity of the masking sound. The degree of masking is indicated by the number of decibels that the hearing threshold is raised owing to the existence of the masking sound.

The explanation commonly given to account for the masking effect is that, if and when a group of nerve fibers is vigorously stimulated by a given sound, that group cannot function effectively to convey other impulses to the brain. In effect, the "lines" are already loaded.

Several important practical results follow from the masking effect. It was implied above that the presence of a background of sound has the effect of raising the threshold of hearing. As a result, a person who is partially deaf may be able to hear as well in a noisy room as one who possesses normal hearing. If the general sound level, in the form of noise, were, say, 50 db, and the threshold of hearing of the semideaf person 50 db, his auditory response would not be materially affected. However, the masking effect of the 50-db noise would, for all practical purposes, raise the threshold of hearing of the person having normal ears to the same level as that of the person having defective hearing. As a result, those who are attempting to carry on a conversation in the presence of a strong background noise usually speak more loudly. Under those circumstances, the partially deaf individual is enabled to hear better than in quiet surroundings.

When a public-address system is used (Sec. 17–2), it is important to keep the ratio of signal (desired sound) to background noise as large as possible, otherwise amplification does not result in a gain in auditory intelligibility—the noise is amplified as much as the desired sound. A broadcasting studio is so designed that the background noise is held at a negligible value, thus giving a high signal to noise ratio, and hence little if any masking effect obtains.

The masking effect may materially modify the quality of musical tones. For instance, if one listens to music over a radio set under noisy room conditions, or in the presence of electrical interference, the character of the musical tones will be decidedly changed. This result will obtain because the weaker tones and the weaker components of complex tones will, to a greater or less extent, be masked by the existing background noise. It is therefore obvious that high-fidelity musical reproduction necessitates the lowest possible background noise.

In preparing the score of a musical composition to be rendered by an orchestra or an organ, the composer must take into account the masking effect of heavy bass passages on the less loud higher tones. Likewise, the artist who plays the composition must keep in mind the possibility that the masking effect may profoundly modify the musical results that he desires to attain.

5–8. *Auditory Fatigue*

When one considers the multitude of sounds that constantly assail the ear, it is surprising that there is not a pronounced dulling of the auditory response. And we are not here referrring to the masking effect just discussed. Rather what we have in mind is the possible effect that a single tone may produce upon the ear if maintained for a prolonged period. By fatigue is meant, in this connection, a transient loss of aural sensitivity due to previous auditory stimulation. Is there such an effect? A considerable amount of study has been given to this subject, and the results of these investigations appear to show that such a phenomenon does exist.

Aural fatigue may manifest itself either by causing an elevation of the hearing threshold or by decreasing the subjective loudness. The factors that influence the magnitude of the fatigue effect are: duration of the stimulating tone, the intensity, and the frequency. In the matter of the elevation of the threshold, the effect of previous stimulation is of relatively short duration, ranging from a few seconds to a few minutes, depending apparently upon the frequency of the exciting tone. The maximum fatigue effect occurs at the frequency of the stimulating sound. Tones below 500 cps produce little effect, but for frequencies above 1000 cycles the effect is quite pronounced, being of the order of 4 db. Surprisingly enough, fatigue effects may be noted when the stimulation is only slightly above the threshold of hearing.

This phenomenon, and the effect to be discussed in the next section, have a bearing on the auditory reception of musical sounds. For a more detailed discussion of fatigue the reader is referred to "Theory of Hearing" by E. G. Wever and "Hearing" by Stevens and Davis. The foregoing observations on the subject of fatigue constitute, to some extent, a résumé of the discussions of this effect to be found in the above-mentioned books. These authors have thoroughly reviewed the available research reports, and their findings are therefore authoritative.

5–9. *Persistence of Sensation*

We are familiar with the persistence of vision. The fact that a retinal image persists for an appreciable interval of time makes

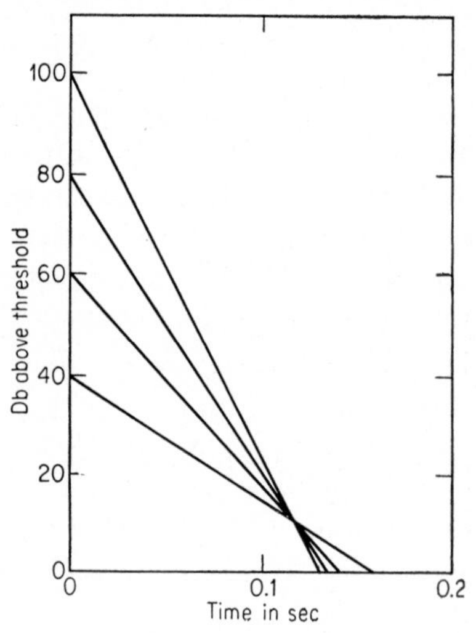

Fig. 5–12. Decay curves for an 800-cycle tone. (Reproduced by permission of John Wiley & Sons, Inc., from "Hearing" by Stevens and Davis, 1938.)

"moving pictures" possible. The "time lag" in the case of optical impressions is of the order of $\frac{1}{50}$ sec. Is there a corresponding auditory phenomenon? The answer is in the affirmative. However, in the case of hearing, the time interval involved is much shorter, and hence quite difficult to determine. Probably the most reliable studies of auditory persistence have been made by Békésy.[1] He found that in the case of a particular frequency the sound reached the threshold of hearing in 0.14 sec regardless of the initial intensity. This is shown by the plots appearing in Fig. 5–12. If further study should show that this decay time is independent of the frequency, and also if there is a corresponding delay in the build-up of sensation, we would have an important auditory constant. This constant would have a bearing on the question of the turning on and off of a tone in such a manner as to avoid a click and yet have it appear that the sound has been started or stopped instantaneously. Further, the persistence of a sensation quite possibly has a bearing on the musical effect of chords in the execution of which the several notes are not struck or released simultaneously. By the same token it is also conceivable that the tonal effect of notes played in rapid succession might be influenced by the sensation time lag. Obviously there is need for additional study of this phenomenon. Here is a tempting problem in the field of psychoacoustics.

QUESTIONS

1. What is meant by the *intensity* of sound, as heard at some point in the sound field?
2. Define the "decibel."
3. What is the difference, in decibels, between the intensity level of two sounds if their pressures are 0.002 and 20 dynes per sq cm respectively?
4. If two sounds differ in intensity levels by 57 db, the intensity of the louder sound is how many times that of the fainter sound?
5. What is the intensity level of a barely audible 100 vps tone?
6. At what levels do the intensity and loudness levels of a 500-cycle note practically coincide?
7. A change in the intensity level from 30 to 60 db produces what change in the loudness level of a 200 vps tone?
8. A change in the loudness level from 30 to 50 db of a 300 vps tone is produced by what change in intensity levels?
9. At what frequency is a barely audible tone produced at 40-db intensity level?
10. Why does the presence of a background noise affect one's ability to hear a relatively faint sound?

[1] G. V. Békésy, Über die Hörsamkeit der Ein-und-Aus-schwingvorgänge mit Berücksichtigung der Raumakustik, *Ann. Physik.*, vol. 16, pp. 844–860, 1933.

6 *Resonance*

6–1. *The Phenomenon of Resonance*

One of the phenomena encountered in the study of sound, particularly in the field of music, is that of resonance. Many sources of sound, such as the vocal cords, a violin string, or the lips of a horn player, would, by and of themselves, produce only faint sounds. It therefore becomes necessary to provide some means whereby the intensity of the weak sounds emitted by such sonorous bodies may be increased, and thus the loudness of the sound augmented. To accomplish this desirable end advantage is taken of the principle of resonance. In order to arrive at an understanding of the nature of resonance and the way in which it is employed in the domain of music, it will be well to consider the matter of resonance in general terms.

If a lath, fastened at one end, has its free end pulled aside and then released, it will oscillate back and forth a certain number of times per second, its period of oscillation depending upon its mass, length, and other mechanical characteristics. A wire fastened at both ends and under tension will, if plucked, continue to describe transverse vibrations for several seconds, and the frequency of its oscillations will be a function of its tension, mass per unit length, etc. In other words, the lath, or the wire, disturbed from a state of rest and subsequently left free to vibrate, will continue to oscillate in its own particular natural period of vibration.

Furthermore, if the exciting impulses are properly timed, exceedingly small increments of energy will cause a body that is capable of oscillating to describe relatively wide oscillations; in other words, to vibrate through a comparatively wide amplitude. This important fact may be shown by means of a very simple experiment. If a mass whose weight is 5 or 10 lb is suspended by a suitable cord from a rigid support, it may be set into oscillation by means of a fine thread which is also fastened to the weight. If one gives a slight and transient pull to the delicate thread, it will impart to the heavy mass a small displacement. If now one continues to give additional slight pulls to the thread **at properly timed intervals,** it is possible to build up a considerable amplitude of oscillation. In order to accomplish this, however, the impulses must be **isochronous;** that is, **they must have the same period as the natural period of the suspended body.** This experiment shows that relatively large masses of matter can be thrown into oscillation by imparting very small amounts of energy, applied periodically.

Another illustration of this phenomenon may be arranged by suspending three pendulums from a common support, preferably a rigid one, as indicated in Fig. 6–1. The two pendulums P_1 and P_3 have the same length and hence the same natural periods of oscillation; P_2 is shorter and hence has a different period. If now P_1 is caused to oscillate, P_2 will not be disturbed, but P_3 will, in turn, begin to swing. The slight amount of energy conveyed periodically through the rigid support is sufficient to bring about the oscillation of P_3, **because $\mathbf{P_3}$ has the same natural period as $\mathbf{P_1}$;** isochronism exists. In the two instances cited we have examples of mechanical resonance; many other cases might be mentioned. In general then, it may be said that mechanical resonance will occur when the period of the exciting pulses of energy coincides with the natural period of the body involved. As we shall see in the discussions that follow, there are many instances in the field of musical acoustics where the principle of resonance plays an important part. In such cases a solid or a gas (usually air) may act as the exciting agent (the generator or driver), and likewise similar types of bodies may function as resonators.

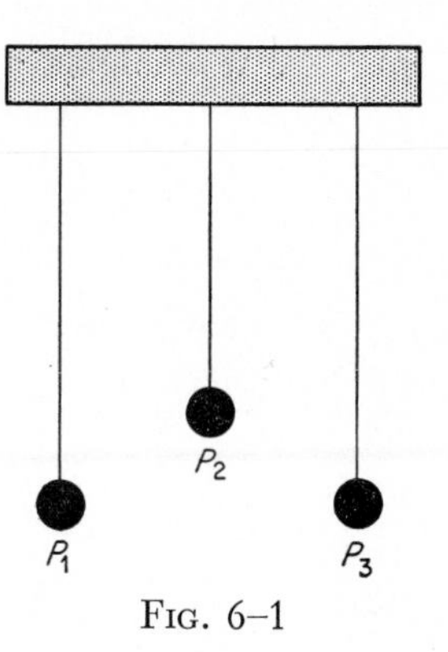

FIG. 6–1

6–2. *Acoustical Resonance*

There are many common instances in which free sound waves act as the source of energy whereby a resonance effect is established. If one holds down the sustaining pedal of a piano, each string becomes a resonator having a definite natural period of vibration. If a note be sung near the instrument, it will be found that one or more of the strings will be set into sympathetic vibration, thus acting as a resonator, and may be heard to emit a tone corresponding to the note sung. In fact several tones may possibly be heard owing to the existence of several components (overtones) in the original sound. The air in rooms, particularly in those of small dimensions, often has one or more natural modes of vibration. By singing notes one can often sense that room resonance occurs as a note of a certain pitch is sung. In the baptistry at the cathedral of Pisa there is a striking example of room resonance. For a small fee the attendant, who has a deep voice, will intone a particular note to which the air in the room will loudly resonate. The matter of room resonance is a subject of considerable importance in connection with architectural acoustics. Under certain circumstances the effect is beneficent, while under other conditions it may prove to be an acoustical nuisance. But in any event, room resonance should not be confused with room reverberation. The latter phenomenon is a matter of multiple reflection, as pointed out in Sec. 3–9.

Another instance of resonance is to be found in the so-called "sound of the sea" that may be heard when one holds a rather large sea shell to the ear. Such a natural acoustical unit probably resonates to several frequencies, and thus the enclosed air is thrown into sympathetic vibration by the various sounds generated by the waves and wind.

Another example of acoustical resonance occurs in a tuning fork. When sounded alone a generator of this type gives rise to a faint sound. If, however, the fork is mounted on a hollow box whose internal dimensions are of such a value that the natural period of the enclosed air is nearly the same as that of the fork, a relatively loud sound will result. The body of enclosed air is thrown into sympathetic vibration owing to the small periodic increments of energy received from the vibrating fork.

A still more striking example of acoustical resonance is that of a heavy tuning fork being set in vibration by sound waves from a second fork. If two forks, accurately adjusted to the same frequency and mounted on resonating bases, are placed several meters apart, one of the forks will respond, because of resonance, when the other is caused to vibrate. Indeed, forks are now made which will manifest such sympathetic vibration when separated a distance of something like 100 ft. When one recalls that the acoustical power in such cases is measured in millionths of a watt, the fact that a second fork can be thrown into sonorous vibration at such a distance by means of incident sound waves serves as a striking confirmation of the resonance principle.

Resonance effects in connection with air columns are of particular significance. This important case is easily illustrated by bringing a vibrating tuning fork near the open end of a column of air of adjustable length, as shown in Fig. 6–2. By pouring water into the jar a definite length of air column may be found that will loudly resonate in response to the isochronous impulses of the fork. Sounding by itself, the fork emits but a very small amount of energy per second. However, when resonance occurs the combination may be heard throughout a good-sized room. This demonstration is illustrative of the basic part played by the resonance principle in connection with the design and operation of a number of musical instruments, particularly those of the wind-blown type. Later we shall return to this application of the resonance effect, and in that connection it will also be noted that the length of such a resonating air column bears a definite relation to the pitch of the generator.

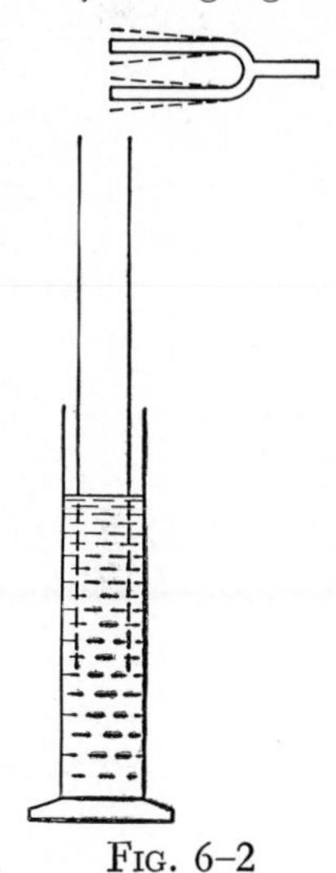

FIG. 6–2

6–3. *Cavity Resonators*

Professor Helmholtz made use of the principle of resonance in the construction of the famous resonators that he utilized in the analysis of complex sounds, and in the detection of sounds that would otherwise be inaudible. These resonators were usually made of spun brass in the form of hollow spheres, as shown in Fig. 6–3. In use, the sound waves entered the opening *a*, while the aperture *b* was inserted in the ear. As originally designed and used by Helmholtz, these resonators were made in different sizes, each designed to respond to a particular frequency. Resonators of the same general character as the Helmholtz units are now commonly made in a cylindrical form and so designed that the volume is adjustable. Figure 6–4 shows a modern form of such an acoustical unit. If the dimensions of the resonator are small compared with the wavelength involved, the resonance is quite sharp. The frequency to which such a cavity resonator responds is determined by its volume and by the size of the aperture through which the sound waves enter.[1]

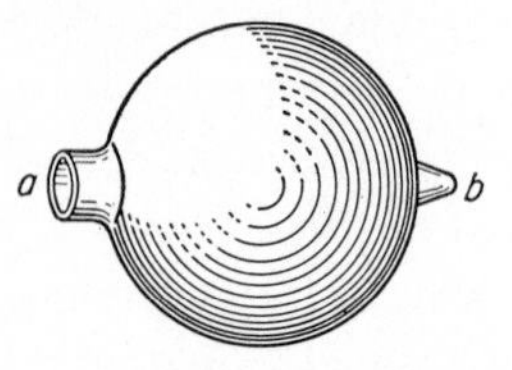

Fig. 6–3. Helmholtz resonator.

Fig. 6–4. Modern form of resonator.

It is interesting to note in passing that the principle of resonance was apparently recognized by the Greeks and the Romans. It is said that they utilized large acoustically tuned vases placed in their

[1] A working formula by which the resonant frequency of such a resonator may be computed is to be found in Knudsen and Harris, "Acoustical Designing in Architecture," p. 123.

theaters for the purpose of augmenting the loudness of certain frequencies considered to be important in the rendition of speech and music.

Cavity resonators are frequently utilized in applied acoustics, two common examples being in connection with the marimba and the organ stop known as the "celestial harp."

6–4. *Absorption of Sound by Resonators*

In the preceding section we have discussed the utilization of cavity resonators in connection with the augmentation of sound. We now turn to a consideration of a reverse effect—the absorption of sound by means of resonators. If a cavity resonator, tuned to the frequency of the source, is introduced into a sound field at a point somewhat distant from the source of sound, it will be found that an appreciable amount of sound energy is absorbed over a portion of the advancing wavefront, with the result that the loudness will be diminished in the region of the resonator. This absorption of energy results from the frictional losses at the aperture as pronounced variations in pressure take place within the resonant cavity. Cavity resonators have been used in buildings for the purpose of absorbing certain frequencies.[1]

In transmitting sound through an acoustical conduit it is sometimes found desirable to suppress one or more components of a complex wave disturbance. This can be accomplished by the use of one or more cavity resonators arranged as shown in Fig. 6–5. If the resonators are all tuned to the same frequency, each will absorb a certain amount of energy from the passing wave train, and thus materially reduce the loudness of that particular component. If the ear is placed at *B*, that frequency will be almost wholly absent from the sound issuing from the tube. By tuning each of the resonators to a different frequency, a band of frequencies can be largely suppressed. An assembly of this general character is known as an **acoustic wave filter.** There are factors other than absorption

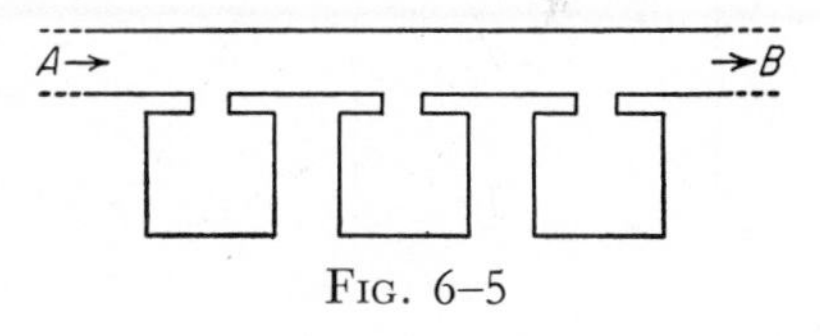

FIG. 6–5

[1] See a paper by V. L. Jordan, *J. Acoust. Soc. Am.*, vol. 19, p. 972, 1947.

involved in such devices. The complete theory of wave filters is more or less complex, and it is beyond the scope of this book to deal with the subject in detail. They are mentioned here only as an illustration of the absorption property of cavity resonators. For an extended discussion of the theory of wave filters the reader is referred to "Acoustics" by Stewart and Lindsay.

6–5. *Broad Resonance*

From what has been said in the preceding section of this chapter, it is not to be inferred that perfect isochronism must obtain in order to bring about resonance. Sympathetic vibration will occur when the frequency of the driver is not exactly that of the resonating body. If the periods of two bodies differ by a few vibrations per second resonance may occur, but the response will be less marked than when perfect synchronism obtains.

There is also the possibility that a sympathetic response may occur when the natural periods of vibration of the generator and the second body are widely different. Reference is here made to the case in which a very broad resonant response occurs. Certain bodies or systems of bodies will often respond to a comparatively wide range of exciting frequencies. This is particularly true when the damping, or dissipation of energy, due to friction is relatively high. This type of resonance is sometimes referred to as general or **broad resonance,** and is illustrated graphically by curve *B* in Fig. 6–6. The dotted line represents the frequency of the generator. Curve *A* represents the response when the frictional damping is low. There are various important applications of this type of resonance, the most notable of which are the action of the sounding board of a piano, and the results which obtain in the instruments of the violin family. In such cases the sounding board, or the body of the instrument and the enclosed air, reinforce a wide range of tones and

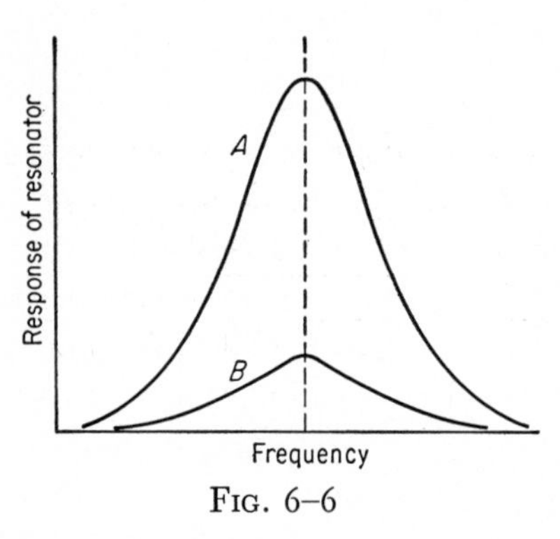

FIG. 6–6

thus serve to augment greatly the sound emitted by the strings themselves.

Speaking broadly, it may be said that the acoustical effectiveness of practically all musical instruments depends upon the principle of resonance. It is thus apparent that an understanding of the nature of resonance, as well as its possibilities and limitations, is of the utmost importance in the study of musical acoustics. In later chapters we shall consider in detail the utilization of this principle in the design and functioning of the various types of musical instruments.

6–6. *Energy Relations in Cases of Resonance*

In concluding our discussion of resonance one other aspect of this important phenomenon should be noted. Reverting, for the moment, to the experiment of the tuning fork and the associated resonant air column, the question at once arises as to whether the increase in loudness due to the presence of the resonator means that the available energy has been increased. In answering this inquiry it must be kept in mind that the freely vibrating fork (the driver) is the only source of energy involved. The resonant air column in the cylinder receives whatever energy it may have from the fork. How then can the increase in loudness be accounted for? An increase in loudness is of course due to an increase in intensity. In Sec. 5–3 intensity was defined in terms of the time rate of energy delivery. If the length of time during which a fork will continue to vibrate with and without the presence of the resonator is noted, it will be found that the fork will cease to vibrate much more quickly when associated with the resonator than when the air column is absent. In other words, the energy content of the vibratile component is more quickly dissipated when a resonator forms a part of the sonorous system. It may then be concluded that **the resonator does not create energy** but that it brings about the emission of energy at an increased rate. Thus the intensity is increased, and hence the subjective loudness is enhanced. So, in general, it may be said that the presence of a resonator increases the radiation efficiency of the generator, and thus augments the loudness. And this relation is involved in connection with practically all musical instruments, including the vocal cords.

QUESTIONS

1. Explain how you would utilize the phenomenon of resonance for the purpose of determining the frequency of a tuning fork.

2. If you were to find the fork referred to in question 1 to have a frequency higher than desired, how would you adjust it to the proper frequency?

3. For what practical purpose are resonators used?

4. Distinguish between "broad" and "sharp" resonance. Cite examples of each type.

7 *Pitch*

7–1. *The Meaning of Pitch*

Musical sounds may differ from one another in three respects, viz., in loudness, in quality, and in pitch. Loudness has been defined and discussed in Sec. 5–5. The subject of quality will be treated in the next chapter. Pitch has to do with the position of a sound in a musical scale, as judged by the listener.

In dealing with the three characteristics mentioned above we shall be considering the psychological aspects of sound, in contradistinction to the objective aspects, such as radiation, frequency, and intensity. The magnitude of such quantities as those just enumerated can be measured by means of instruments. But in the cases of loudness, quality, and pitch, we are dealing with subjective responses. The magnitude of such reactions cannot be quantitatively determined because their evaluation depends upon individual judgment.

In considering the subject of simple harmonic motion in Chap. 2, it was found convenient to use the term frequency as being synonymous with pitch. Actually the terms do not mean the same thing. Frequency refers to the number of vibrations per second made by the sonorous body; it also indicates the number of oscillations per second occurring in the transmitting medium. Pitch is that subjective characteristic of a sound that enables us to classify a sound as being acute or grave. Obviously there is a multitude of specific values of pitch lying between those two extremes. Pitch and frequency are, of course, intimately related. If the frequency of vibration of a sonorous body is high, the pitch, as sensed by the listener, will be high; if the frequency is low, the pitch will be low. It is known that our estimate of pitch is, to some extent, influenced by the loudness level of the sound involved. This is particularly true in the case of pure tones at comparatively low frequencies. Because of this fact it has been suggested that the pitch of any given tone should be expressed as the frequency of a pure tone whose loudness level has a specified value, for instance 60 phons. Facilities are available, based ultimately on an accurate fork, or group of such forks, whereby a series of pure tones may be caused to reach loudness level of a standard value, say 60 phons. The pitch of an unknown tone can then be determined by adjusting the frequency of the standard source until a skilled auditor or group of listeners decides that the two tones, when sounded alternately, have the same pitch. The frequency of the variable standard would then be the pitch of the tone under study.

As indicated above, it has been found that in the case of tones whose waveform is sinusoidal the pitch of a sound appears to fall as the loudness level increases, particularly in the lower frequency range. In the range between 200 and 300 cps if one increases the stimulation from, say, 60 to 100 phons the pitch may drop 10 per cent. Fletcher found, for instance, in the case of a pure tone emitted by a sonorous body whose frequency was 261, the pitch dropped 1.3 semitones when the loudness level was increased from 40 to 80 phons. But in the range over which the ear is most sensitive (1000 to 5000) there is little change in pitch as the loudness level is raised. Furthermore, in the case of the complex tones found in musical notes, intensity and waveform have only a slight effect on pitch. This question is, therefore, largely academic, at least so far as applied music is concerned, though it might have a slight bearing on the matter of dissonance and consonance (Sec. 9–2).[1]

There is, however, another aspect of the subject of pitch that calls for comment. As we shall see later, practically all musical sounds are complex in character, being made up of a so-called fundamental tone (lowest pitch) and a group of other components (overtones) each having a pitch higher than the fundamental. The question that at once presents itself is this: Which one of these components determines the pitch of the tone as a whole? In Fig. 8–15 is to be

[1] For an extended discussion of pitch see Stevens and Davis, "Hearing," chap. 3.

found a diagram that depicts the spectrum of a particular voice. The length of the vertical lines indicates the relative intensity of the several components. It will be noted that the fourth overtone is much stronger than the fundamental. Is it the strongest component that determines the pitch of the tone as a whole or is it the fundamental, even though this component is relatively weak? The answer is that, in general, it is the pitch of the lowest component (fundamental) that fixes the pitch of the tone as a whole. However, it has been found that if a tone is made up of a group of components whose frequencies differ by a like amount, as 200, 300, 400, and 500, the hearer may judge the pitch to be not 200, but 100 cps. The reason for this unusual situation will be discussed in connection with the question of subjective tones (Sec. 7–5).

7–2. *Standards of Musical Pitch*

The basic pitch (A above middle C) employed in musical literature has varied widely since Père Mersenne, the famous French ecclesiast and mathematician,[1] first determined the pitch of a musical note. In Mersenne's time (1648) the lowest "church" pitch was 373.7 and the "chamber" pitch was 402.9. Handel's standard pitch in 1751 was 422.5. At one time in this country a so-called "concert" pitch of 461.6 was used. Probably the first highly accurate determination of the pitch of a sonorous body was made by Lissajous, another eminent French physicist. Lissajous determined the frequency of the standard tuning fork of France, known as the "diapason normal," of the French Conservatory of Music. It was intended that this standard fork should execute 435 vibrations per second, but a later determination by the famous acoustician, Rudolph Koenig, with improved facilities, showed that the diapason normal actually gave a frequency of 435.45 at 15°C or 59°F. Probably the first formal action to adopt a standard pitch occurred in Germany when a meeting of physicists at Stuttgart in 1834 adopted a pitch of 440.

[1] Mersenne is sometimes referred to as the Father of Acoustics. His most important work entitled "Harmonie universelle," in twelve volumes, appeared in 1636. The founding of the French Academy of Science is credited to Father Mersenne.

An orchestral A of 435 was legalized in France in 1859, and this pitch was soon adopted by several important symphony and opera orchestras, including the Boston Orchestra (1883). In 1892 the Piano Manufacturer's Association adopted the French pitch of 435, as determined by Koenig, and designated that value as "international pitch."

In 1939 an international conference on pitch was held in London, and it was unanimously agreed to recommend to all interested organizations that 440 be adopted as the standard of orchestral pitch. This pitch is now generally used in this country. In order to assist in maintaining an accurate pitch value, the Bureau of Standards at Washington, D.C., broadcasts several times daily a standardizing pitch of 440 by means of radio signals.

The basis of pitch in orchestral renditions is the second open string of the violin, designated in musical terminology as A_4. The violins check their A pitch against the A of the oboe player, who in turn takes the pitch from an accurately adjusted tuning fork. It would probably be better if orchestras and bands were to make use of an electronic pitch-standardizing device so designed that all players in the orchestra could hear the primary standard. It is entirely possible to design and construct a small portable pitch-emitting unit embodying an accurate temperature-compensated tuning fork, with associated amplifier and small loudspeaker, that would give a constant pitch under practically any ambient conditions. Several symphony orchestras have recently been considering the adoption of such an innovation.

It may be well to note in passing that there has long existed what is commonly known as "philosophical pitch." Early in his acoustical work Koenig adopted a series of vibration frequencies, probably based on a suggestion said to have been originally made by the acoustician Sauveur, and later used by Chladni. This series of pitch values is built up by taking for the octaves successive power of 2. On this basis $C_4 = 2^8 = 256$, $C_5 = 2^9 = 512$, etc. This would give to A a value of 426.6, which is close to the A of Handel. Since the beautifully made and highly accurate forks made by Koenig came to be widely used in scientific laboratories, it was natural that physicists came to use the so-called scientific, or philosophical, pitch ($C = 256$). There is, however, no good reason for continuing the use of such a pitch value. Now that an international pitch has been adopted, physicists and acoustical engineers should discard the scientific pitch (A = 426.6) and officially adopt the value 440 now in common use throughout the musical world.

7–3. *Change of Pitch with Motion—the Doppler Effect*

In discussing the subject of pitch, and its relation to frequency, mention should be made of a phenomenon sometimes observed, particularly in these days of high-speed travel. Many will have noticed the apparent change in pitch which occurs when one is rapidly passing a railroad signal bell, for instance. As one approaches the source of sound the pitch is appreciably higher than normal and when we move away from the sounding bell the pitch is observed to be lower. The same phenomenon occurs when a source of sound is moving and the observer is stationary, and also when both the observer and the source are moving, with respect to one another. In each of these cases the change in pitch is due to the relative motion of the observer and the source.

The phenomenon above referred to is one instance of a more general principle first noted by Doppler in 1842. It is not difficult to account for this effect. Keeping in mind that the term pitch signifies the position in a scale (Sec. 7–1), to which we assign a musical sound, and also remembering that it is the number of sound waves per second which strike the ear which determines pitch, we may arrive at an understanding of the Doppler effect by examining the diagram shown in Fig. 7–1. We will assume that a source of sound is stationary and located at S, and that an observer is moving from E toward E' with a velocity v'. Let the velocity of the sound be v. If both the source and the observer were at rest the observer at E would receive n waves per second, n being the frequency of the vibrating body as a source of sound. The frequency will be given by the simple relation $n = v/\lambda$, where λ is the wavelength. If, however, the observer moves from E to E', **more waves will reach his ear per second,** while motion is taking place, than would be the

Fig. 7–1

case if no relative motion obtained. He will receive the original number of waves per second, **plus** the number of waves between E and E'. This additional number will be determined by the relation v'/λ. Hence the total number of waves reaching the observer per unit of time will be given by the expression

$$n + n' = \frac{v}{\lambda} + \frac{v'}{\lambda} = \frac{v + v'}{\lambda}$$

Under these circumstances, then, it will be evident that the pitch, as sensed by the observer, will be higher than the normal pitch of the sounding body, and by an amount determined by the velocity of the observer. Here is a case where frequency and the pitch number are not numerically equal.

In a similar manner, relations may be worked out for those cases where the source moves, and where both the source and the observer move. From the equation given above, and from the other relations just referred to, it is evident that if the relative velocity of the source and observer is constant, the pitch will change by a fixed amount. If, however, the relative velocity is not constant, the pitch also will change. The latter situation is often noted when two automobiles pass, one of which is sounding the horn and the other is changing speed as they pass. The effect which we have been discussing is a matter of relative velocity; the distance between the sounding body and the observer is not a factor in the case.

7–4. *Aural Acuteness*

Reference has already been made (Sec. 5–6) to the general pitch limitations of the normal ear, and it was there pointed out that the ear is capable of responding to an exceedingly small amount of incident energy. It is interesting and of practical significance to learn something about the minimum pitch difference which the ear is capable of recognizing. A skillful piano tuner can distinguish between a true and an equally tempered (Sec. 9–7) fifth. This involves the recognition of the one-fiftieth of a semitone. This means that the tuner is called upon to recognize not less than six hundred different sounds in an octave. Indeed, it is said that there is evidence for believing that, in the region of middle C and under favorable conditions, it is possible to distinguish one note from another when the pitch differential is only 1/120 of a semitone. While such marked acuteness is to a great extent a matter of training and experience, it is surprising to note how small a pitch differential can be recognized by even the untrained ear. Dr. Knudsen, of the University of California at Los Angeles, has investigated this problem, and his findings are shown graphically in Fig. 7–2.

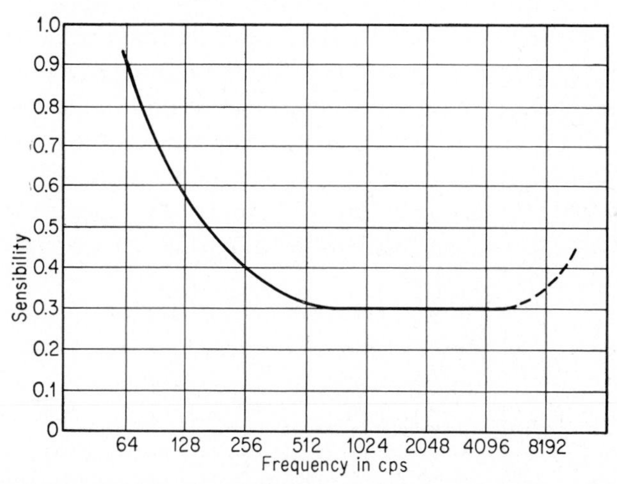

Fig. 7–2. Minimum Perceptible Differency in Frequency, in terms of $\Delta f/f$. (Reprinted by permission from "Architectural Acoustics" by Knudsen, published by John Wiley & Sons, Inc.)

As would be expected from our study of intensity sensitivity (Sec. 5–6), it is found that the ear is less sensitive to pitch differences at the lower frequencies. The graph shows that the maximum pitch sensitivity falls in the region between 500 and 4000 cycles. In that band of frequencies the ear can distinguish a difference in pitch as small as 0.3 per cent, or less than a sixteenth of a semitone. However, at 60 cycles, which is approximately the lower limit of the bassoon, the smallest change in pitch which the ear can detect is of the order of 1 per cent; this is a sixth of a semitone. The region above 5000 cycles has not yet been thoroughly investigated, but it is known that the pitch sensitivity falls off sharply at the higher frequencies, probably in some such way as indicated by the dotted portion of the curve. The fact that the ear behaves, functionally, as indicated above should receive consideration in connection with musical composition and rendition, though it probably is not given proper recognition at present.

Another aspect of pitch perception which is coming to have increased significance concerns the minimum time required for the ear to recognize a definite pitch. The data bearing on this point are, as yet, not very concordant, but it would appear that the **"pitch perception time"** is more or less independent of the frequency, being of the order of one-twentieth of a second.

Not only can the ear catalogue a note as to pitch in a remarkably short time interval, but it can also recognize a sound as such when only two vibrations are made. This means that in the region of C_4 we can become conscious of the existence of a sound if it persists for less than 1/100th of a second. However, to recognize the complete characteristics of a note anywhere from 2 to 20 oscillations are necessary.

7–5. *Subjective Tones*

In considering the matter of pitch it should be recognized that the hearing mechanism may so react to a given objective stimulus that one may sense a frequency **which is not present in the original sound wave.** In other words, the ear may, in effect, create new frequencies out of the incident disturbances.

In order to understand the reasons for this strange auditory reaction it is necessary to examine the behavior of a vibratile body when subject to a periodic disturbance. If we take a simple undamped diaphragm supported at its periphery and cause a train of sound waves, for instance, to fall upon it the diaphragm will be set into vibration and the to-and-fro motion will be symmetrical with respect to its resting position. In such a case the response is said to be **linear,** i.e., the displacement of the diaphragm is strictly proportional to the excitation. In terms of symbols this fact could be expressed by

$$R = kE$$

when R represents response, E magnitude of excitation, and k a constant which involves the mass and the elasticity of the particular diaphragm being used. In the case of sound waves the excitation *E* could, for instance, be expressed in terms of pressure variations.

If, however, our diaphragm is in some way mechanically damped, as the result of being in contact on one side with a more or less flexible body, the response of the vibratile member will be considerably different than in the simple case first cited. The chances are that under these conditions the to-and-fro motion of the diaphragm will not be symmetrical with respect to its resting position. In such a case the response is said to be **nonlinear.** As a result of this unsymmetrical response of the diaphragm the wave motion which is conveyed to the articulating parts is not a replica of the original waves which served to excite the vibratile component. In other words, though the incident wave motion may have been sinusoidal in character, the resulting vibration of the diaphragm (and associated parts) will not be; but instead will be of a complex nature.

It may be shown by mathematical analysis that a sound wave which has undergone distortion because of the nonlinear action of a receiving agent has acquired various overtones. The particular overtones thus introduced will depend upon the receiving agent and upon the acoustical circumstances involved. As a matter of fact, wave distortion may occur if the intensity of the exciting wave disturbance is great, even though there is little or no damping. In other words a receiver, such as the ear mechanism, may show a linear response for sounds of low intensity (of the order of a few db) but exhibit a nonlinear response for loud sounds, say 80 db or more. This means that the ear not only transmits the original frequency but may also introduce various and sundry other frequencies with the result that the nerve impulses reaching the brain centers contain disturbances which had no objective genesis. For instance, if one were to amplify the pure tone of a tuning fork until the sound had an intensity level of the order of 80 or 90 db, the auditor would probably hear a number of overtones, perhaps a complete series. Or if one were to sound some musical tone which contained in addition to the fundamental only the odd numbered partials, the listener might also be conscious of at least some of the even partials. Thus, because of the nonlinear response of the tympanum or other components of the ear mechanism, various **subjective tones** may be introduced into the final auditory result. Whether such subjective tones are present, in any given case, will depend upon the intensity of the sound, as well as upon the particular listener. In the field of music, as we shall see later, the possibility of the appearance of subjective tones is of considerable importance. Subjective tones are sometimes referred to as **aural harmonics.**

There is still another group of phenomena which results from the nonlinear action of the hearing organization. If two fairly loud musical tones of different pitch be sounded simultaneously, the auditor may hear a third tone whose pitch differs from either of the original tones. Musical history records that Sorge, a German organist, in 1745, and Tartini, an eminent Italian violinist, in 1754 independently discovered this phenomenon. Tartini called such tones "terzi suoni" or "third sounds." They are often referred to as **Tartini's tones.** As a result of modern research we now know that one may be conscious of a whole series of such tones—tones which are not present in either of the original sounds. Indeed the exciting sounds may both be pure, i.e., entirely free from overtones, and yet a listener will testify that he hears one or more such pseudotones. We are here dealing with another form of subjective tones.

The pitch relationship of such subjective tones to the generating tones can best be illustrated by citing a simple example. Suppose we amplify the sound emitted by two tuning forks having frequencies of, say, f_1 (300) and f_2 (500) cps. Analysis and experiment show that the listener may hear one or more of a series of subjective tones having the following pitch relationships:

$$
\begin{aligned}
2f_1 &= 600 \\
2f_2 &= 1000 \\
f_2 - f_1 &= 200 \\
f_1 + f_2 &= 800
\end{aligned}
$$

The first two frequencies in the above list will be recognized as the octaves (first overtone) of the respective original tones. These are to be expected from our previous discussion; but the last two frequencies apparently bear no simple relation to the primary frequencies. The subjective tone whose frequency is given by the difference between the primary frequencies is usually referred to as a **difference tone** and the tone whose frequency is the sum of the original frequencies is spoken of as a **summation tone.**[1] Collectively, difference tones and summation tones are frequently referred to as **combination tones.** The difference tone is not difficult to hear, and is probably the one Tartini observed. Summation tones are not so easily recognized. However, if the primary tones have a relatively low pitch, thus making the frequency of the summation tone fall near the middle of the audible range, these tones may be heard, particularly if the original tones are loud, thus causing the ear to function in a nonlinear manner.

Because the frequency of difference tones would correspond to the frequency of beats it was originally felt that such tones were what might be called *beat tones*, but the consensus of opinion at present is that difference tones are quite distinct from any beat phenomena; and it is obvious that summation tones cannot be explained on a beat-tone basis. It is now believed that both difference and summation tones result from the nonlinear response of one or more components of the ear structure. Under certain circumstances additional upper partials and other difference and summation tones than those given in the above list may be heard.

Under suitable circumstances it is possible for a listener to be conscious of a somewhat different type of difference tone than that just described. Reference is here made to a subjective tone which functions as the fundamental of a group of overtones.

If, for instance, a tone having a fundamental frequency of 220 is sounded on a trombone we have present, as will be shown later (Sec. 13–18), not only the fundamental but a complement of five upper partials. The frequencies of these overtones, in such a case, would be 440, 660, 880, 1100, and 1320. Now it will be observed that the difference between any two adjacent partials is 220, which is the frequency of the fundamental tone. If one now, by means of a suitable acoustic filter, removes the fundamental from the trombone tone (a procedure which can readily be carried out), the listener will still hear the fundamental, though it has been entirely suppressed. The difference tone supplied by one or more pairs of the existing upper partials **subjectively takes the place of the actual fundamental.** In short, the ear supplies the fundamental. Indeed one may suppress the fundamental and one or more partials in the

[1] The discovery of summation tones is ascribed to Helmholtz.

original tone and the ear itself will supply these missing tone components. Here we have a case where the ear in some way develops a subjective tone from existing upper partials, while in the case previously cited overtones were created, subjectively, from one or more existing fundamentals. It has been established that the nonlinear response of the ear is responsible for this type of difference tone.

The practical significance of subjective tones is to be noted. If partials are absent in the original tone they may be added by the ear, and then the quality of the tone (Sec. 8–1) as heard may be materially changed. If the fundamental is absent or weak in a given tone this component may be introduced or strengthened by the ear.

A striking example of this is to be found in the response characteristics of small loudspeakers which form a part of some radio receiving sets. Many loudspeakers have a cone diameter of six inches or less. It is probable that, in most cases, such a cone will not respond effectively to frequencies much below middle C (262), and yet we hear at least some of the bass notes of an orchestral rendition. The tones of many of the strings and horns are rich in harmonics. From these components the ear creates, as difference tones, the missing fundamentals and probably some of the lower partials. This does not mean that the net result is as satisfactory from a musical point of view as if the speaker itself reproduced the lower notes, but the subjectively created lower notes may, at least in some cases, serve as acceptable substitutes for objectively produced sounds.

Differential tones are utilized in various ways. One example of this is to be found in the design of the whistles often used by game referees and policemen. These small units consist of two very short pipes having a common supply of air. The pipes are slightly unequal in length and hence when simultaneously sounded emit two tones that differ somewhat in pitch. Under these circumstances a difference tone is heard whose pitch is much lower than that of either of the pipes constituting the combination. It is thus possible to make a whistle having very small compass but yielding a tone of comparatively low pitch.

Organ designers have sometimes made use of difference tones in order to avoid the use of the long pipes needed to produce the very

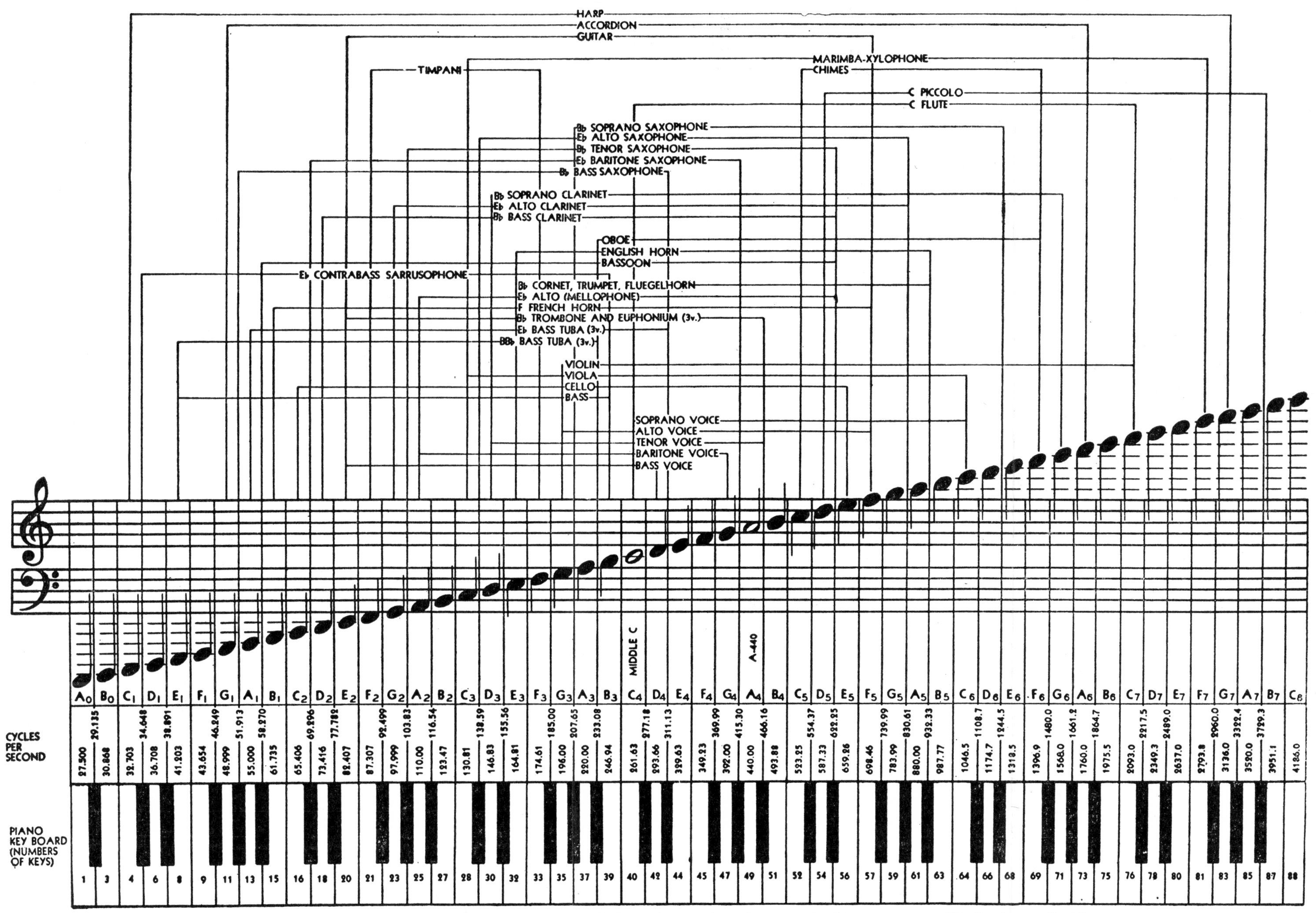

Fig. 7–3. Chart showing the frequency range of the voice and various musical instruments. (C. G. Conn, Ltd.)

low organ tones. In such cases the resulting combination tone is sometimes referred to as an *acoustic base*.

In a later chapter having to do with the quality of musical sounds we shall again refer to combination tones. Chap. 7 of "Hearing" by Stevens and Davis should be consulted for additional details on aural harmonics and combination tones.

7–6. *Pitch Range of Voice and Instruments*

The average singing voice has a compass of about two octaves, though in a few cases it may exceed this range. For instance there is a soprano (Yma Sumac, a Peruvian), now engaged in concert work, who is said to have a possible range of five octaves, and a usable range of four full octaves. The ordinary compass attained by male and female singers is indicated on the accompanying chart (Fig. 7–3). The diagram shows that the combined range of male and female voices totals about three and one-half octaves (82–1046). It will be apparent that the average vocal soloist employs about twenty-five different notes.

Of the musical instruments the organ has the widest pitch range, extending from 16.4 vps, given by the 32-ft pipe, to 8372 sounded by a pipe whose speaking length is ¾ in. Some of the larger organs have a range of eight octaves, but usually the span is seven octaves or less.

The highest note employed in orchestral music is given by the piccolo, which has a pitch of $A\#_7$ (3729.3). The lowest orchestral tone is that of the bass viol, E_1 (41.2). The highest piano note is C_8 (4186), and the lowest A_0 (27.5). The ordinary pitches of the voice and of the more common musical instruments are indicated in the chart shown as Fig. 7–3.

7–7. *Production of Standard Frequencies and Measurement of Pitch*

In acoustical studies it becomes necessary to generate sounds over a wide range of frequencies that will serve as standard pitch values. There must also be available some practical method whereby one may make accurate comparisons between the pitch of any given sonorous body and that of an accepted standard. Let us first examine the means by which one may generate tones having an accurately known pitch. Probably the first laboratory method of determining pitch involved some form of a siren. Such a device consists of a metal disk in which are several concentric series of holes. The disk is caused to rotate at a known angular speed. When a jet of air is directed against the holes, a simple tone is produced, and the frequency of the tone can be computed from the number of holes in a given series and the angular speed. The comparison is made when a listener judges the siren-generated pitch to be equal to the pitch of the source under test. This method of determining pitch is now largely obsolete, though it is still sometimes used, at least as a check method.

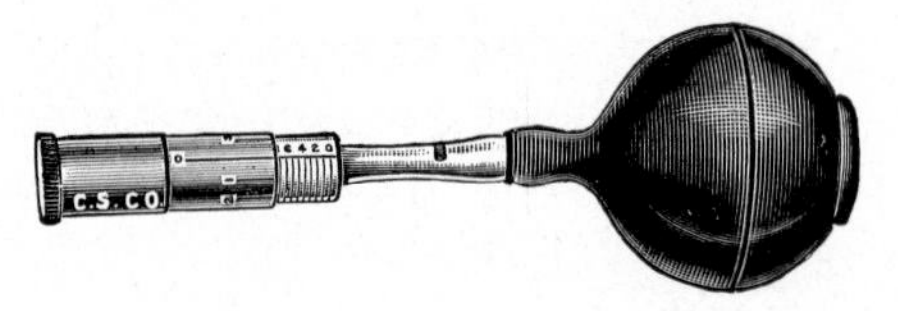

FIG. 7–4. Galton's whistle. (Central Scientific Co.)

A classical device, originally designed by Captain Douglas Galton, is sometimes still used as a means of producing high-pitched tones of known value. An illustration of a Galton whistle is to be seen in Fig. 7–4. Air from a rubber bulb is forced into a small metallic whistle having a calibrated chamber. Sounds of known pitch up to the highest audible limit can be produced by this simple device. Comparison with the standard can be made by auditory judgment or by the use of the beat method.

In the time of Koenig accurately made tuning forks constituted the accepted method of generating sounds of known frequency, particularly in the higher frequency bands, and the forks made by Koenig himself have never been surpassed for accuracy. He constructed and calibrated a series of such generators, the upper limit of which was 90,000 vps (λ = 1.9 mm or 0.075 in.)—more than two octaves above audibility. The frequency of any given tone could be determined by comparing it with one or more of the standard forks by the method of beats. This consists of adjusting the frequency under test until no beats occur when the standard sources and the unknown are sounded simultaneously, this procedure being referred to as a **zero beat method.**

At present forks are seldom made to function at frequencies above about 4000 vps. Beyond this limit bars are used, of the type shown in Fig. 7–5. Such bars are now available ranging in frequency from 4000 to 20,000 cps. They are made of a light alloy, and when rigidly clamped at their mid-point, and struck on the end by a rubber hammer, will vibrate longitudinally, and for a much longer time than high-frequency forks.

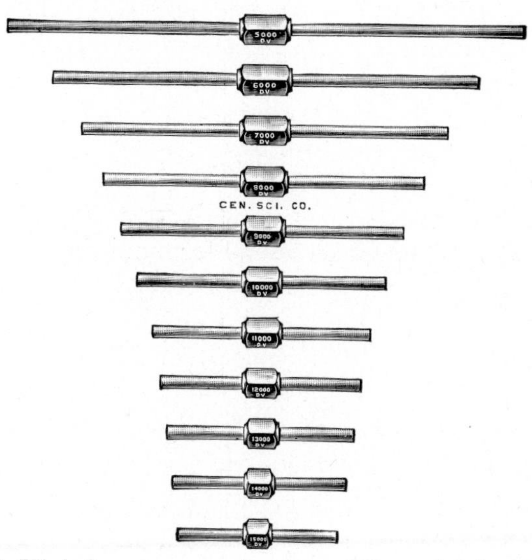

Fig. 7–5. High-frequency tuning bars. (Central Scientific Co.)

A piece of equipment that is now widely used for the production of pure tones, varying from the lowest to the highest, is known technically as a **beat-frequency oscillator,** more commonly referred to as an **audio oscillator.** An audio oscillator consists, essentially, of an electrical organization composed of two electronic circuits so designed that each generates a high-frequency alternating current of the order of 100 kilocycles (100,000 cycles). One of the generators operates at a fixed frequency, while the other is variable. If and when the output from these two a-c generators is fed into a suitable common electrical network, a third alternating current comes into being, and this third alternating current will have a frequency which equals the difference between the two primary electrical frequencies. In other words, we have a case of electrical beats—hence the name of the device. This third electrical current, after being passed through certain amplifying and filter circuits, is fed to a loudspeaker. Thus by varying the frequency of one of the generators, one may secure a beat note of any desired frequency between, say, 0 and 40,000 cycles; and this beat note closely approximates a pure tone. In the illustration the knob on the calibrated dial controls the variable generator, and hence the output: a scale on the edge of the dial indicates the frequency being generated. By varying the degree of amplification faint or loud tones may be produced at will. Sounds of unknown frequency may be compared with the tones produced by the above-described audio-frequency generator, by the beat method. By the use of this piece of apparatus (audio oscillator) it is also possible to determine pitch by the procedure outlined in Sec. 7–1.

Though audio oscillators are useful for many frequency determinations, their calibration is not accurate enough for use in the tuning of pianos and organs. Many tuners, in such cases, rely on a well-calibrated fork or series of forks, making use of a beat method that will be touched upon in the next chapter.

For the accurate tuning of musical instruments, there is now coming into use a very ingenious piece of equipment that makes it possible to determine pitch by an optical method, rather than by the use of an auditory procedure. By this method the personal equation is largely eliminated from frequency determination tests. This relatively new technique makes use of what is referred to as a stroboscopic phenomenon, and the equipment is known under the trade name Stroboconn.[1]

The stroboscopic principle is widely used in timing procedures. If, for instance, a rotating wheel is being illuminated by an intermittent light whose frequency of interruption is equal to the number of times per second a mark on the wheel passes the line of sight, the wheel will appear to be at rest. An example of such a phenomenon is sometimes observed when looking at motion pictures; the wheels of a moving vehicle appear to stand still, or to be rotating backward. Motion-picture film is projected at the rate of 24 frames

[1] A somewhat less elaborate unit is known as a Strobotuner.

per second, and because of the "optical lag" of the eye, the successive pictures blend into an optical impression of motion. But if it so chances that a wagon wheel is rotating at such an angular speed that 24 spokes pass any given point each second, the eye no longer senses motion and the wheel appears not to be rotating. If, however, fewer than 24 spokes pass the reference line per second the wheel will appear to be turning slowly backward. By the same token, a faster angular speed will give to the wheel an appearance of a forward motion. The motion appears to be stationary only when the rate at which the frames are projected exactly equals the product

FIG. 7–6. Sector-disk used in the Stroboconn. (C. G. Conn, Ltd.)

of the number of spokes multiplied by the number of revolutions the wheel makes per second.

Another example of the application of the stroboscopic principle is to be found in the adjustment of the angular speed of a phonograph turntable. A sectored paper disk is placed on the turntable and illuminated by means of a neon bulb. When supplied by the ordinary 60-cycle alternating current the lamp will flash 120 times per second. The markings on the disk are so arranged that when the turntable is rotating at the desired angular speed the black sectors will appear to remain stationary.

The Stroboconn is so designed that the frequency of illumination given by a neon lamp is caused to vary as the frequency of the sound varies. Stroboscopic sectors (Fig. 7–6) are caused to rotate at a constant angular velocity, the speed of the driving motor being very accurately controlled by a tuning fork. There is a series of seven segments on each disk, and there are 12 such rotating disks, thus providing for a total of 84 possible pitch elements. Such an arrangement provides for a setting on each note in a span of nearly seven octaves. The sound from the instrument is picked up by a microphone and converted into an a-c current. This current, properly

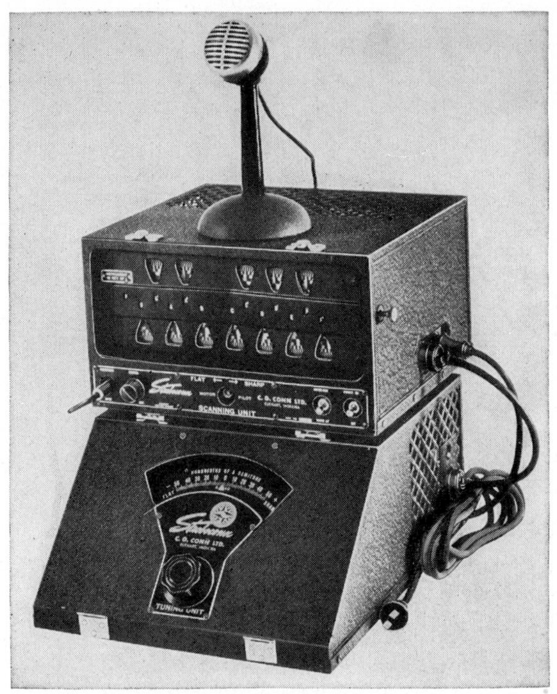

Fig. 7–7. Stroboconn. (C. G. Conn, Ltd.)

amplified, causes a neon lamp to flash at a frequency corresponding to the pitch of the tone being studied. In operation some one segment ring may appear to be stationary, thus indicating the pitch of the sound incident on the microphone. If, however, a ring pattern rotates slowly to the left, it means that the sound has a frequency somewhat lower than the standard; if the rotation is to the right, the tone has a higher frequency than the standard. Provision is made so that the standard frequency may be set at 440 cps, or at any other value. From the foregoing brief description of the Stroboconn it will be seen that this apparatus indicates whether a tone is flatter or sharper than the standard, and by how much. The divergence can be read in hundredths of a semitone (cents), and with an accuracy greater than is possible by an aural method. Figure 7–7 is an illustration of the Stroboconn assembly. By the use of this piece of equipment it is possible to determine frequency values accurately and rapidly.

QUESTIONS

1. What is the significance of the term *pitch*, and how does this concept differ from the idea of *frequency?*
2. What is the generally accepted standard of musical pitch?
3. If you had no secondary standard of pitch, how would you check the accuracy of your primary standard fork?
4. A person is driving at a speed of 75 ft/sec toward a source of sound, a siren for instance, whose pitch is known to be 400 cps. What is the pitch that the person actually hears, assuming the speed of sound to be 1100 ft/sec?
5. Many of the loudspeakers on small radio sets do not reproduce sounds whose frequencies are below 150 cycles, yet one seems to hear the bass notes of a musical composition. How do you account for this?
6. How does the pitch range of the flute and the piccolo compare?
7. If the pitch of an organ pipe is correct when the temperature is 70°F will it be higher or lower if the room warms to 80°F?
8. Why is it important that an orchestra tune just before the program begins, rather than earlier?

8 *Quality*

8–1. *Nature and Cause of Quality*

Having considered two of the characteristics of a musical sound, we next proceed to a study of the third and most important aspect, quality. By **quality** is meant that characteristic of a sound by which one is able to distinguish it from all other sounds of like pitch and loudness. We readily recognize the different voices of our acquaintances; the sound of a violin can be distinguished from that of a trombone, even though the two instruments play the same note at the same intensity level; and the difference in the nature of the sound generated by two different makes of violins is recognizable. The quality or character of the sound arising in each of these cases results in a definite and particular sound impression. The French use the word **timbre** to express this characteristic of a sound; and the Germans also have a word, **Klangfarbe,** a free translation of which is "tone color," which is used by them to designate what we term quality.

Not only do our ears tell us that different sounds have different characteristics, but the graphic representations of the corresponding sound waves also show that definite differences exist. In Fig. 8–1 we see the recorded waveforms of the sounds emitted by three well-known sources. In making these recordings the pitch and loudness level was held approximately equal in the three cases. The marked difference in waveform indicates, then, that some factor or factors, other than pitch and loudness, give rise to the difference in tone character that we refer to as quality or timbre.

Our first problem is to arrive at an understanding of the objective cause of what we have termed quality. Why does one sound, par-

Fig. 8–1. Waveforms of tuning fork (*top*), clarinet, and cornet, each at a frequency of 440, and at approximately the same intensity.

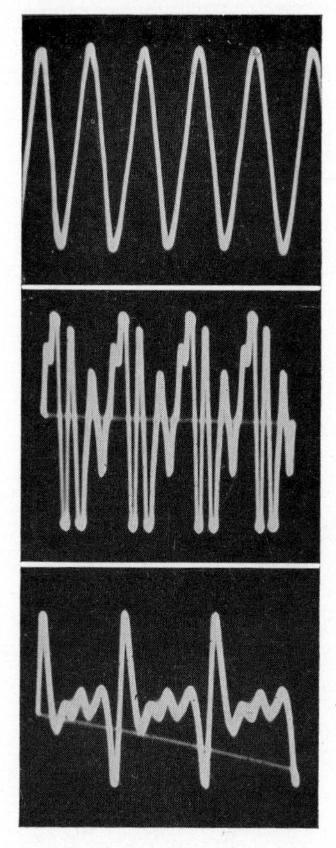

ticularly a musical sound, differ in character from another, similar, sound?

We have agreed that a muscial sound consists of a periodic motion in some medium, usually air. This means that the motion of the particles constituting the medium repeats itself once during each single period, and this regardless of the character of the motion. It is thus evident that, even though we hold the pitch and intensity constant, a wide variety of motions might give rise to a sound. It was Prof. G. S. Ohm, a German physicist and mathematician, who first pointed out the physical basis of quality. According to Ohm the **motion** of the particles of the transmitting medium corresponding to a composite musical sound is in reality the sum of a group of **simple** periodic motions; and for each such simple oscillation there exists a simple tone, of definite pitch, which the ear can detect. It accordingly follows that all but simple (pure) tones are composite. The several components which go to make up such a complex sound structure are called **partial tones,** or briefly, **partials;** the partial having the lowest frequency is designated as the **fundamental.** The partials having frequencies higher than the fundamental are referred to as **upper** partials or **overtones.** In many instances the frequencies of these overtones are exact multiples of that of the fundamental; and in such cases the fundamental and the upper partials are, together, called **harmonics.** In those cases where the frequencies of the overtones are not exact multiples of the fundamental, the elemental tones are indicated by the term **inharmonic** partials.

The diagram appearing as Fig. 8–2 shows the fundamentals and upper partials of the tones emitted by the three instruments whose waveforms are depicted in Fig. 8–1. Such charts are known as sound spectra. The length of the vertical lines indicates the relative strength of the several harmonics.

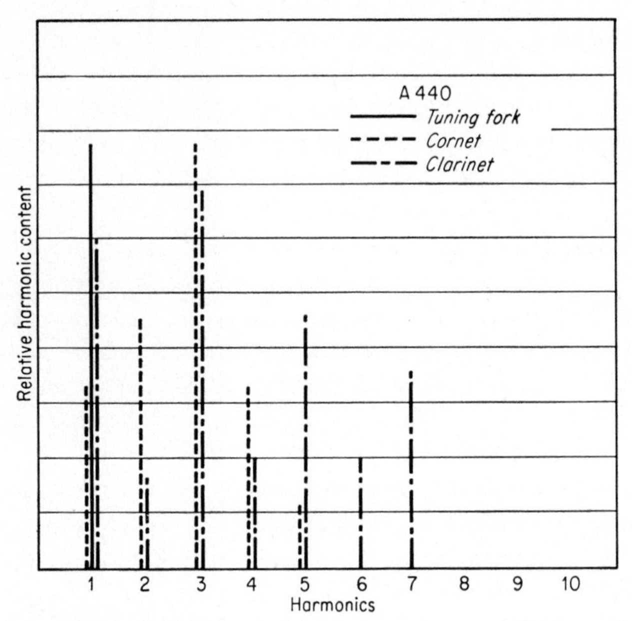

FIG. 8–2. Spectra of instruments referred to in Fig. 8–1.

Helmholtz carried out a long and carefully devised series of experiments for the purpose of testing the validity of Ohm's law of acoustics, as it has come to be called. In 1862 Helmholtz published a volume entitled "Die Lehre von den Tonempfindungen" in which was embodied the result of eight years of study and research in the field of the theory of musical sounds. An able English translation of this classical work has been made by A. J. Ellis under the title "Sensations of Tone." Helmholtz summarizes his findings with regard to quality in these words:[1] "Hence we are able to lay down the important law that **differences in musical quality of**

[1] H. von Helmholtz (Ellis transl.), "Sensations of Tone," 5th edition, chap. 6, p. 127.

tone depend solely on the presence and strength of partial tones, and in no respect on the differences in phase under which these partial tones enter into composition."

Various investigators since Helmholtz's time have confirmed that part of his statement which is emphasized above. There is, however, a serious question concerning the validity of that part of his conclusions having to do with phase. As able an investigator as Koenig takes issue with Helmholtz on this point. As a result of his researches Koenig was led to the conclusion that differences of phase cannot be neglected when accounting for differences in tonal quality. Lloyd and Agnew, of the U.S. Bureau of Standards (1909), investigated this matter and concluded that the phase differences of the components do not produce an appreciable effect on the quality of a tone. Recently, however, W. L. Barrows, of M.I.T., in reporting on the design and operation of a new sound-generating device, states: "It has been observed that a variation of phase not only makes striking differences in the multitone waveform as seen on the oscillograph, but that it also produces marked differences in the character of the sound." While it should be said that the sound studied by Barrows was not one made up of harmonic partials, yet these and other findings would appear to indicate that phase difference may be a factor in the determination of quality. For instance, Firestone and his co-workers have shown (see *J. Acoust. Soc. Am.*, vol. 5, p. 173, 1934; also vol. 9, p. 24, 1937) that a change in phase relationship may, under certain conditions, give rise to a perceptible change in both loudness and timbre.

In considering the factors that enter as determinants of quality it is to be noted that there is experimental evidence for believing that a decided change in the intensity of a sound may produce an appreciable change in timbre. It should also be added, in this connection, that the quality of a musical sound may be modified by the existence of aural harmonics (Sec. 7–5).

With the foregoing reservations in mind, one might summarize the situation by saying that the quality or timbre of a musical sound is determined, **chiefly,** by the number, intensity, and distribution of the partials that enter into its composition.

The first sixteen harmonics, based on C_2 as a fundamental, are shown in the diagram appearing as Fig. 8–3. A corresponding harmonic series might be set up on a fundamental having any given frequency. Such a series does not end with the sixteenth partial; indeed, the series may extend into the ultra-audio region. The numerals in the first line below the musical staff indicate the **order** of the harmonics; in the second line are given the frequencies of the several harmonics; and the last line shows the frequency of the notes on the equally tempered scale (Sec. 9–7). It will be noted that the harmonic partials are not exactly in tune with the notes on the musical scale, except in the case of the octave relationship. The musical notation indicates the scale tones which most nearly approximate the frequency of the true harmonics.

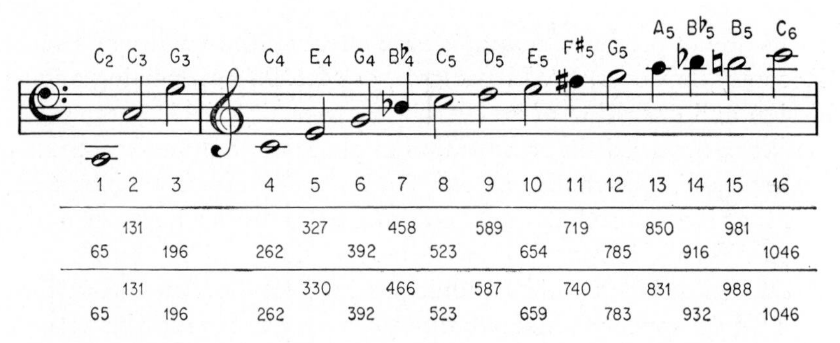

FIG. 8–3. Harmonic series, based on C_2 as a fundamental. Frequencies are given to the nearest whole vibration.

In this connection it is to be noted that there are instances where the frequency of one or more upper partials may not be exact multiples of the fundamental. If the discrepancy is not more than a few cycles the quality of the tone will not be seriously impaired. If, however, the departure from being an exact multiple is appreciable such an overtone constitutes an inharmonic partial, and the resultant complex tone becomes "rough" and hence unpleasant. Inharmonic partials, in general, have relatively high frequencies.

In considering the subject of quality it should be recalled that in Sec. 5–2 it was pointed out that the elements of the inner ear respond to definite frequencies. As a result of this remarkable ability the ear can break down a complex sound into its several components and thus function as a wave analyzer. The neural reactions to the individual frequencies are in turn integrated into a resultant auditory impression that we recognize as having definite tonal characteristics, thus making it possible for the auditor to recognize the sounds generated by different voices and by the various musical instruments. This aural process of analysis and final integration constitutes one of the most remarkable phenomena to be found in all nature. The process bears some resemblance to the tricolor response of the human eye.

In discussing the subject of quality it should be noted that the timbre of a musical sound may undergo a decided change within a small fraction of a second after the vibrating element of the instrument has been excited. This is particularly true in those cases

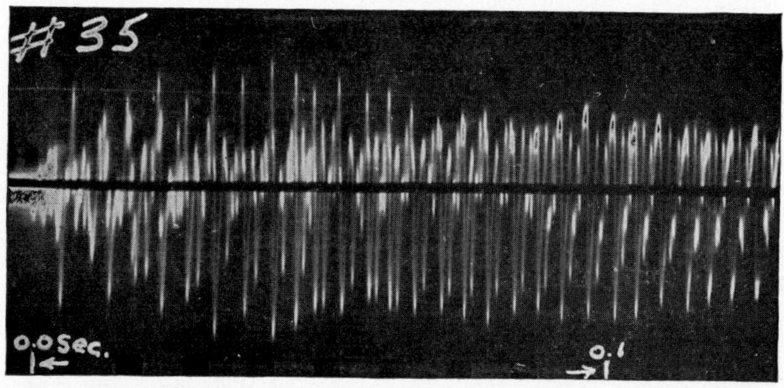

FIG. 8–4. Record showing initial transients in the case of a percussive tone. Piano tone, C_4.

where the sound is initiated by a percussive stroke, as in the case of the piano, the kettledrum, and the xylophone. In such instances the vibratile member is excited by being struck a more or less sharp blow at a definite point in its structure. Under such circumstances, immediately after the stroke, the upper partials, in general, are found to be relatively strong. But owing to the natural damping of the vibrating member or to the damping caused by the hammer while it is in momentary contact with the string or bar, these higher partials tend to decrease rapidly in amplitude, the most pronounced change occurring in less than one-tenth of a second after the initiation of the sound. (Partials which diminish in intensity in this manner are often referred to as **transients.**) As a result of the fading out of these high-frequency partials the character of the sound undergoes an appreciable change. The recordings that are shown in Figs. 8–4 and 14–8 experimentally confirm the foregoing statement. It is therefore apparent that the timbre of what might be called the "strike tone" may be different from that of the "continuing tone."

What has just been said with reference to those cases where a percussive means of excitation is employed also applies, to a greater or less extent, to those instruments in which the strings are excited by being plucked, such as the harp, the guitar, and at times the violin.

Thus we see that the presence of transients is characteristic of the sound emitted by certain instruments, and that this initial group of higher partials, to some extent at least, gives color to the sound emitted by a piano and similar instruments.

At this point a review of the analytical treatment of harmonic motion by Fourier, as outlined in Chap. 2, will be found profitable.

8–2. *Effect of Direction of Radiation*

Before concluding our discussion of the subject of quality, it will be in order to consider the effect of the direction of wave propagation on the character of the sound as it issues from a given musical instrument. Later we shall study in detail the timbre of the sounds radiated by the various musical instruments, but in those cases it will be assumed that the position of auditor is on a line coinciding with the axis of symmetry of the instrument such as P_1 in Fig. 8–5. In practice, however, the auditor may, in some instances, be located at a point corresponding to P_2 or P_3. If the pitch of the sound and its loudness level be held constant, will the sound spectrum be the same at these three representative points? The answer is that the tone quality of a musical instrument will vary with the orientation of the listener with respect to the axis of symmetry of the instrument. This fact is strikingly shown by the variation in waveform as shown in Fig. 8–6. A similar phenomenon is encountered in connection with the radiation from practically all musical instruments,

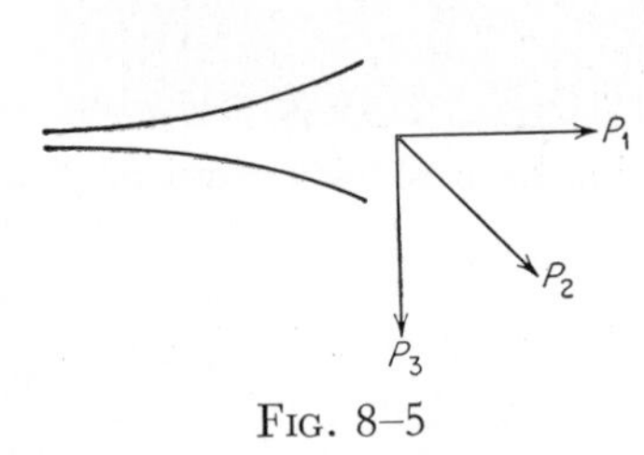

FIG. 8–5

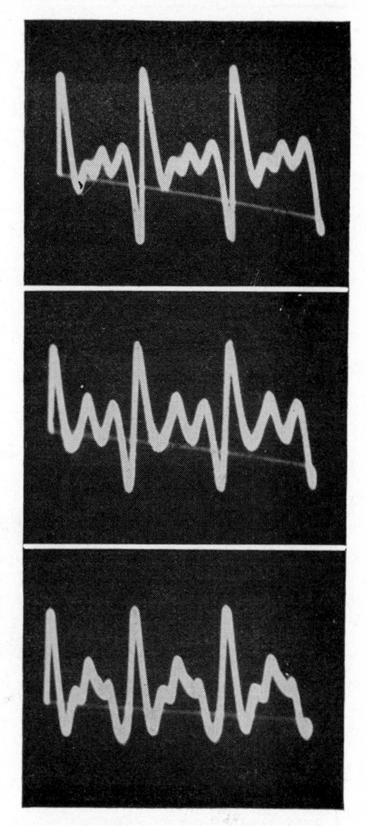

Fig. 8–6. Waveforms of a cornet tone (440) when received on the axis of the instrument (*top*), at 45°, and at 90°. Intensity, and distance from the bell of the horn, held constant.

including the vocal organs. Investigations show that it is the components in the region above about 2000 cps that fall off most sharply as one departs from the axis of symmetry. This is particularly true of the violin, even when considering frequencies as low as 1000 cps. The timbre of a violin is quite different when one listens on the right and then on the left side of the player. All of this has a distinct bearing on the tonal results to be expected from an instrument when the listener is not directly in line with the instrument being played. Band and orchestral conductors might well consider this aspect of the situation when they are assigning the position of the various players.

8–3. *Methods of Observing and Recording Waveforms*

In 1909 the late Dr. Dayton C. Miller, Professor of Physics at Case School of Applied Science, disclosed a method whereby excellent recordings of sound-wave patterns could be made. Professor Miller made use of an original device which he named the phonodeik.[1]

As shown in Fig. 8–7, the assembly consists of a small thin glass disk *D* held between rubber rings, and mechanically connected to an accurately mounted spindle by means of a delicate silk fiber. The spindle carries a tiny mirror *M* that serves to reflect an incident beam of light onto a moving photographic film *F*. Any vibration of the disk

[1] D. C. Miller, The Phonodeik, *Phys. Rev.*, vol. 28, p. 151, 1909.

thus results in a rotation of the pivoted spindle and a consequent deflection of the reflected light beam. Since the motion of the film is at right angles to the displacement of the beam of light, a wavy photographic trace results; thus a trace of the waveform of the incident sound is photographically recorded.

It is to be noted that, essentially, the basic principle of the phonodeik is the same as that suggested in Sec. 2–3, and sketched in Fig. 2–3. However, Miller's apparatus is infinitely more sensitive than the crude assembly there outlined.

By the use of this beautiful piece of equipment, Dr. Miller (himself a skilled musician) made recordings of the waveforms generated

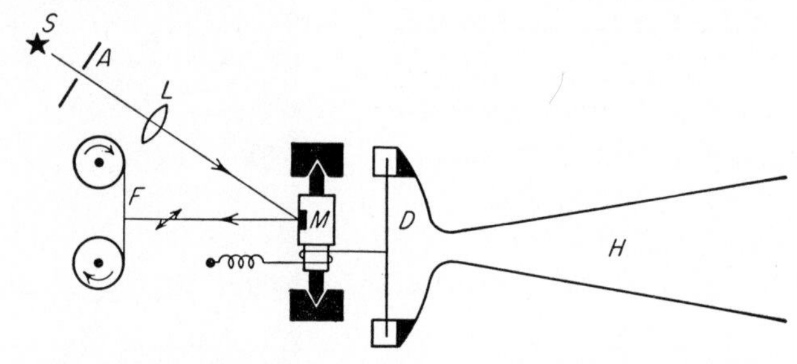

Fig. 8–7. Diagrammatic sketch showing the optical and mechanical components of the phonodeik.

by a number of musical instruments. These recordings have perhaps never been surpassed in clarity and accuracy. Reproductions of some of Professor Miller's original recordings are to be found in his classical volume entitled "The Science of Musical Sounds," to which the reader is referred.

In more recent times other methods of recording and studying sound waves have been developed. These newer techniques, making use of electrical equipment, are more convenient and more rapid than the procedure originally employed by Miller. The author has made a large number of waveform recordings, some of which are used as illustrations in this book, by means of a form of oscillograph. Such an assembly consists of a permanent magnet *NS* (Fig. 8–8) in the field of which is suspended a delicate loop of wire through which the audio-generated alternating current passes. The motion of the loop is mechanically damped by being surrounded by a light transparent oil; electromechanical resonance is thus avoided.

In use the sound waves serve to actuate a microphone, thus giving rise to a weak alternating current which after being amplified passes through the loop above referred to. The alternating field set up by this sound-generated variable current reacts with the permanent magnetic field, thus causing the loop of wire to vibrate. A tiny mirror is attached to the loop and is thereby caused to oscillate in response to the sound waves incident on the mirror. An optical setup similar to that used in connection with the phonodeik is used for making a photographic record.

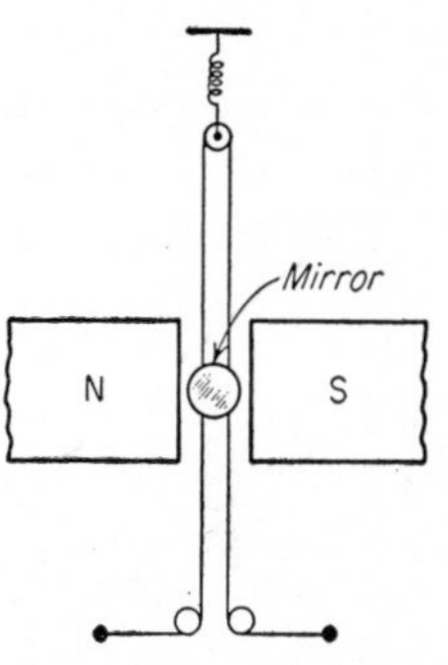

FIG. 8–8. Essential components of electromagnetic oscillograph.

There are certain mechanical limitations involved in the response characteristics of both the phonodeik and the electromagnetic oscillograph as sound-recording devices. One such limitation is due to the fact that the moving parts have appreciable mass. This difficulty is avoided by employing an assembly that makes use of a stream of electrons, bits of negative electricity. For the purposes of this discussion electrons are without mass. Such a device, known as a **cathode-ray oscillograph,** has now largely replaced the earlier equipment utilized in acoustical studies.

The basic component of this modern form of oscillograph consists of a so-called cathode-ray tube, as sketched in Fig. 8–9. Within

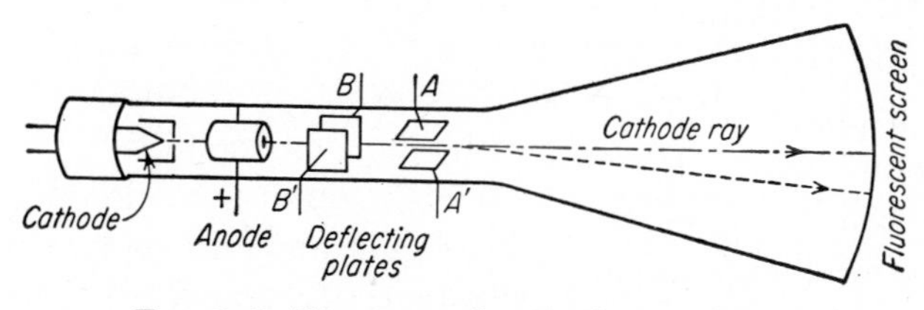

FIG. 8–9. Diagram of cathode-ray tube.

an evacuated tube an electrically heated filament (called the cathode) emits electrons (cathode particles). By connecting a high electrical potential difference between the cathode and another electrode (the anode) these liberated electrons are caused to move in a straight line at a very high speed, as indicated by the broken line labeled "cathode ray." This stream of electrons strikes a chemically prepared screen at the broad end of the tube. This screen is of such a nature that it fluoresces when struck by electrons. (A television tube is a special form of cathode-ray unit, and operates on the same principles as the tube used in an oscillograph.) Since electrons act as charged bodies they can be deflected from their original path by arranging to have them pass near another charged body, much as bits of paper will be attracted to an electrified comb. On their way from the cathode (filament) to the sensitive screen, the electrons are caused to pass between two small parallel plates, A and A', which are connected to the output of an electronic amplifier. The sound waves to be studied are converted into a variable electrical potential by means of a suitable microphone, the output of which is connected to the input of the amplifier. The output of this amplifier is connected to the deflection plates, A and A', of the cathode-ray tube. The general arrangement of the several components is shown by the block diagram, Fig. 8–10. The alternating electrical potential thus generated by the sound waves under test acts upon the electron beam causing it to move up and down through a distance proportional to the changing value of the sound pressure. A second pair of deflection plates, B and B', whose planes are at right angles to the first set, are connected to a suitably arranged source of alternating electrical potential (the so-called sweep circuit), which, if acting alone, would cause the electron beam to oscillate back and forth in a plane at right angles to the first mentioned deflection. When sound waves, then, are incident upon the microphone, the electron beam will be subjected simultaneously to two motions, these two motions being at right angles to one another. The net result is that the fluorescent spot will be caused to trace a curve on the screen which is a replica of the sound-wave form under examination. Thus it is possible to cause a beam of electrons to draw a transient picture of the waveform of any sound wave—a diagram that may be visually observed, or photographed if a permanent record is desired. The response of the electron beam is instantaneous,

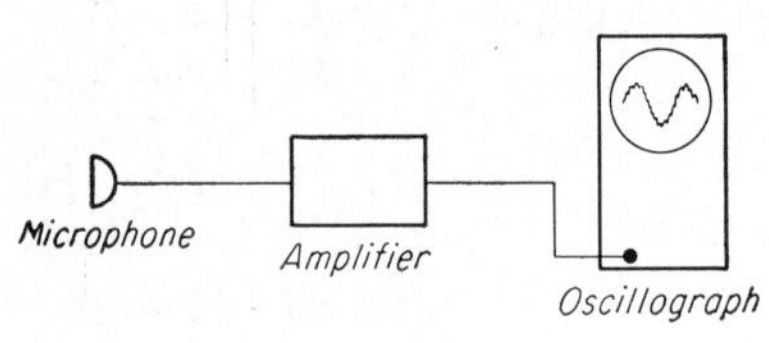

FIG. 8–10. Schematic diagram of cathode-ray assembly

and, given a high-fidelity amplifier and suitable microphone pickup, sound waves of any frequency may be rapidly observed or recorded by this form of oscillograph. In the field of musical acoustics the cathode-ray oscillograph has proved to be a research tool of great value. Figure 8–11 gives a view of a representative model of such an assembly. The circular area in the upper part of the illustration is the screen end of the cathode-ray tube. The knobs on the panel serve to effect control of the image size and position. The waveforms shown in Fig. 8–6, and many other, similar, illustrations in this book, were made by means of such equipment.

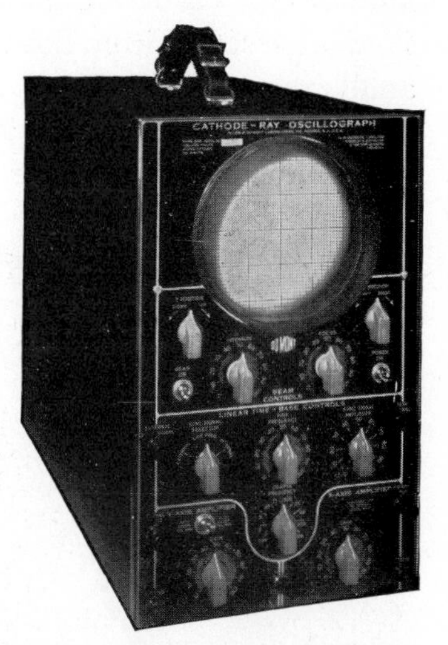

FIG. 8–11. Cathode-ray oscillograph. (Allen B. Dumont Laboratories, Inc.)

8–4. *Sound Analysis*

In the fields of musical acoustics and communication engineering it is important not only to observe the waveform but also to determine what components are present in a given sound, and the relative strength of such components. Such an end is accomplished by means of a piece of equipment known as a **wave analyzer.**

In his classical research work on sound Helmholtz made use of a series of special resonators that bear his name (Fig. 6–3). With these he was able to determine what particular harmonics were present in a given complex sound. However, the use of such resonators does not give a quantitative measure of the relative strength of the partials, and in modern acoustical study this factor is particularly important.

A number of mechanical wave analyzers have been devised for use in connection with the determination of the harmonic content of musical sounds. Such devices, commonly called **harmonic analyzers,** are more or less complicated in character, and considerable time is required to make a complete analysis. Notwithstanding these limitations analyzers of this type have been widely employed in sound analysis. A sound curve, such as those shown in Fig. 8–6, is passed into the machine and, by the manipulation of certain component parts, the relative intensity of the several harmonics can be determined. The sound charts appearing in Fig. 11–5 were set up from data secured by means of such a mechanical analyzer.

One form of a modern wave analyzer consists of a complicated electrical network of thermionic tubes and circuits which makes it possible not only to determine what harmonics are present but also to read on a meter the percentage of each partial present in the original sound. A photograph of a standard analyzer of this type is reproduced as Fig. 8–12.

Fig. 8–12. Wave analyzer. (General Radio Co.)

It is beyond the scope of this volume to enter into the details of the electrical circuits used in the modern wave analyzer, but it may be said, briefly, that the sound under investigation is picked up by some form of microphone, converted into a corresponding alternating electrical current, and then passed into the analyzer. When the proper adjustments have been made the frequency dial is rotated until a meter reading shows. When the meter reading is a maximum, the corresponding dial reading will give the frequency of the particular harmonic that has been singled out. When the controls are suitably adjusted the meter reading will indicate the relative strength of that particular component. After the reading of one component is noted, another harmonic is sought for and a reading made of it, etc. In using the apparatus, the sound being analyzed must persist at a constant level long enough to enable the operator to make a complete series of settings. In the case of a piano, or even with a wind-blown instrument, it is often difficult to make a complete analysis before there is an unavoidable change in the input level.

Various attempts have been made to devise an analyzer that would give **simultaneous** readings of all harmonics present in a given sound. One such unit is referred to as a *panoramic spectrum analyzer*. When this assembly is used, the sound being studied is picked up by a microphone, as in the analyzer just described, and,

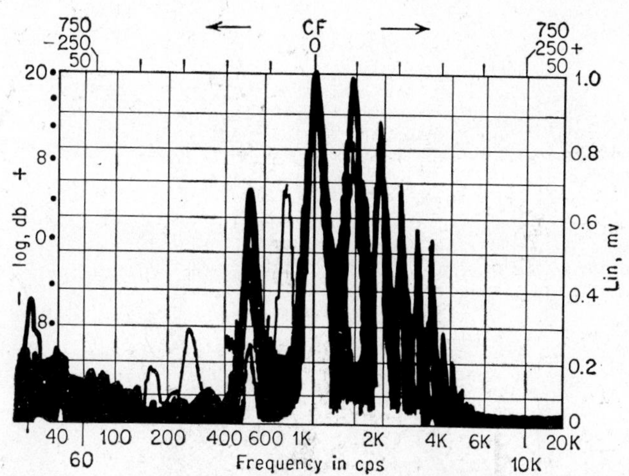

FIG. 8–13. Wave trace made by panoramic analyzer. Trumpet tone, frequency of fundamental 500 cps. The scale at the right indicates the relative magnitude of the several harmonics on a linear basis. The ordinates at the left show the relative magnitudes in terms of decibels. The scale at the top has no significance in this connection. (Panoramic Radio Products, Inc.)

after amplification of the electrical currents thus developed, the electrical energy is fed into an electronic network the output of which serves to actuate an indicator in the form of a cathode-ray tube. The circuits are so arranged that, instead of showing the waveform on the fluorescent screen, a series of vertical, pointed deflections are displayed, one deflection for each harmonic present. The height of each deflection indicates the relative magnitude of the component. Figure 8–13 is a photographic record showing the nature of the image on the screen when a sound is being analyzed. The functioning of this type of analyzer is such that it is necessary that the sound under study must persist at a constant level for at least one second. In addition to this limitation there is another disadvantage in connection with the operation of this type of analyzer. The position of each harmonic along the horizontal scale is on a logarithmic basis; hence the frequency readings are crowded together at the higher frequency end of the display.

FIG. 8–14. Harmonic analyzer developed by the author for use in the study of musical sounds. The meter readings indicate the several harmonics present in a particular tenor voice. The singer was intoning the syllable *ah* on A 220.

The author has designed an analyzer that will, within certain limits, avoid the disadvantages noted in connection with the above-mentioned units. In this analyzer the sound to be analyzed is transformed into a weak alternating current by means of a calibrated microphone. After being amplified somewhat, this current, which is an electrical replica of the original sound wave, is passed into a specially designed electrical filter system which serves to break down the complex waveform into its harmonic components. Each component which may be present is then further amplified. A meter in the output circuit of each channel indicates, in percentages, the relative intensity level of the respective harmonic components. The meter readings can be visually observed or photographically recorded. As now constructed, the new analyzer is designed to give

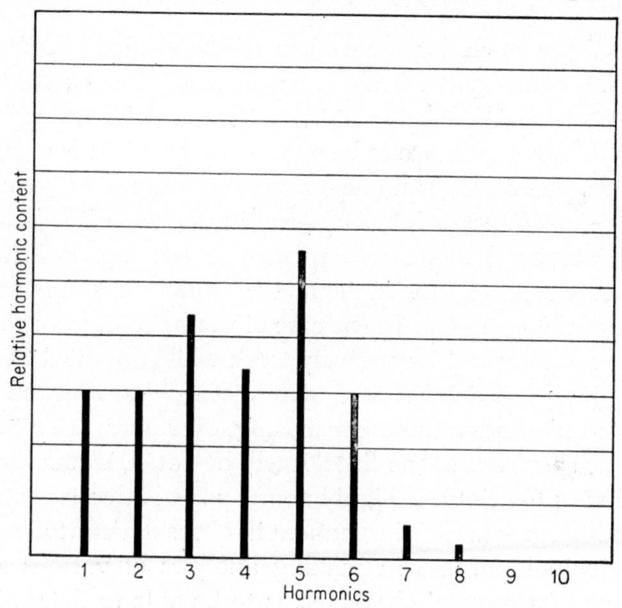

FIG. 8–15. Spectrum of voice referred to in Fig. 8–14.

the first ten harmonics based on two fundamentals whose frequencies are 220 and 440. Harmonics beyond the tenth modify only slightly the quality of a given tone, but the range of the analyzer could be extended, if desired. By noting the readings as given by the several meters, one can quickly determine the character of any musical sound based on the note A below middle C or on a note an octave higher (440). By thus analyzing two notes an octave apart, one can quickly determine the general tonal characteristics of most musical instruments. A photograph of the new analyzer appears as Fig. 8–14. The meter readings shown give the harmonic content of the sound of a tenor voice. Figure 8–15 shows the spectrum of this voice. This analyzer is particularly useful in connection with the study of musical sounds which cannot be held at a fixed intensity level. A number of the sound-spectrum diagrams appearing in this book were made by the use of this form of analyzer. An article describing, in semipopular language, the author's analyzer appeared in the April, 1951, issue of the magazine *Symphony*.

8–5. *Sound Spectra and Their Use*

The analyses made by some form of the equipment described above are commonly set forth in the form of charts called sound spectra, to which reference has already been made. Several typical graphs of this character are to be seen in Chap. 13. It is to be noted that vertical distances (ordinates) represent amplitude (usually expressed in percentage or intensity level in db) while the horizontal distances (abscissas) indicate frequencies, often expressed in terms of the harmonic series. Notice the marked differences in the general character of these spectra. In the case of the bassoon (a reed instrument) the fundamental is relatively weak while the third harmonic is the dominant partial. With the trumpet we find that the fundamental and the next three succeeding overtones are unusually strong. The spectrum of the flute, it will be noted, is comparatively simple; that of the violin is highly complex. It is these spectral differences which give to each instrument its characteristic tone quality.

The principal purpose in making analyses of various sounds, particularly of a musical character, is to be able to determine the cause of changes in quality as one modifies the construction or operation of a given sonorous body. For instance, what is there about the construction of a particular type of piano which makes possible the production of acceptable or even superior tonal results? The same might be asked about wind instruments. Until recently any improvements in design were largely the result of skill in workmanship based on experience plus a "cut and try" method of experimentation. Now it is possible to approach problems of this character from an engineering point of view; today the acoustician can make **quantitative** measurements; and such procedure, if followed, could lead to marked advances in the quality of musical instruments.

8–6. *Synthesis of Musical Sounds*

Thus far in our study of the quality of sounds, we have had recourse to the analytical method only. Following that procedure, we have found that most sounds are complex in character; that is, they are composed of a group of simple tones, the one having the lowest pitch being called the prime, or fundamental, and the others being referred to as overtones, or upper partials. We have seen that the presence of these partial tones makes it possible to distinguish from one another the sounds emitted by various sonorous bodies.

The question now arises: Having analyzed a complex sound, can one reconstruct or **synthesize** such a sound? If this can be done, the theory of sound quality would receive striking confirmation. Fortunately the answer is in the affirmative. In fact, Helmholtz[1] did this very thing. He designed a series of electrically driven forks (each associated with a suitable resonator) giving B♭ (below middle C) as a prime, together with the first seven harmonic upper partials of that note. Mechanical arrangements were provided whereby any one or all of the resonator apertures could be opened and closed at will. By means of this assembly Helmholtz was able to combine any one of the upper partials, or all of them, with the fundamental tone. The intensity of each component was controlled by the same mechanism employed for opening and closing the apertures. In Fig. 8–16 is seen a Helmholtz synthesizing assembly as constructed by Koenig. The Koenig organization contained ten harmonic forks, beginning with $C_3 = 131$. By the use of such apparatus it was found possible to synthesize a sound which has previously been analyzed, and thus confirm the conclusions of Helmholtz regarding the physical basis of sound quality.

Though the compounding equipment made by Koenig was superbly constructed, the scheme of synthesizing devised by Helmholtz had certain inherent limitations; limitations which, owing to modern advances in electrical and related fields, it is now possible to overcome.

At present several methods exist whereby it is possible to produce electrically a tone which is almost completely pure; that is, reasonably free from harmonic content.

[1] Helmholtz, *op. cit.*

For use in certain research studies in connection with the synthesis of musical sounds, the author has designed an electrostatic generator[1] that yields sinusoidal electric waveforms. A group of such generators mounted on a common shaft serve to develop the electrical equivalent of a fundamental and its first seven harmonic overtones. By changing the speed of the motor which drives the alternator shaft, the frequency of the fundamental and its complement of overtones can readily be shifted up or down the musical scale. Provision is made whereby the partials can be singly or

FIG. 8–16. Helmholtz synthesizing apparatus.

collectively fed to a mixing amplifier and thence to a suitable loudspeaker. The amplitude of each partial is also under control. A cathode-ray oscillograph may be connected to the output of the amplifier in parallel with the loudspeaker, thus making it possible both to hear and to observe visually the results of the tone synthesis. Having available the sound spectrum of a given tone, it is possible with this apparatus (called the **Synthephone**[2]) to reconstruct faith-

[1] C. A. Culver, Electrostatic Aternator, *Physics*, vol. 2, pp. 448–456, September, 1932.

[2] For a brief description of this apparatus see a paper in the November, 1939, issue of the *Rev. Sci. Instr.*, by the author.

fully the original sound; and also to create entirely new musical sounds.

It is also possible to construct an assembly whereby sinusoidal components may be produced without the use of mechanically moving parts. For instance, by making use of electron tubes as basic components, one can develop sinusoidal electric currents of any desired audio frequency. Means can be provided whereby the energy output from each oscillator can be controlled. The electrical output from a series of such audio oscillators can be fed into a mixing amplifier and thence into a cathode-ray oscillograph or a loudspeaker. Thus, knowing the audio spectrum of any given tone, the sound can be reconstructed; in short, it can be synthesized.

Various uses are made of such synthesizing equipment, the most notable instance being in the design of so-called electronic organs. In Chap. 15 we shall consider such types of musical instruments.

8–7. *Vibrato*

Before leaving the subject of tone quality, there is one additional determinant which should be mentioned. Reference is made to a melodic embellishment commonly designated as vibrato. This effect is to be observed most frequently in connection with vocal renditions and in the playing of stringed instruments, particularly in the case of the violin. The effect consists of a periodic variation of the tone, the frequency of the fluctuations being two to five per second. Extensive studies carried on by the late Professor Seashore at the University of Iowa have shown that the vibrato is essentially a frequency-modulation effect. There is also evidence to the effect that there sometimes exists a simultaneous amplitude-modulation effect. When both the frequency and the amplitude modulations occur concurrently, the waveform will likewise undergo a periodic change. In the case of the human voice, the vibrato effect appears to be largely involuntary, and possibly may be associated with a periodic lessening of the muscular tension required to produce a sustained note. The emotional content of the passage seems to have some bearing on the degree of the vibrato effect. In the case of a violin or cello rendition, the vibrato effect is deliberately produced by the player as he brings about a periodic alteration of the length of the vibrating string by the movement of his finger. In this case, then, the effect is one that is almost wholly a frequency-modulation phenomenon.[1]

The question arises as to the bearing of vibrato on musical quality. It appears to be a matter of common opinion that a tone in which vibrato occurs—at least in the case of the members of the violin family and the voice—is more pleasing than one that is sustained at a constant level of pitch and amplitude. The effect on the auditor is, to some extent at least, a psychoacoustical one. In any event, the reaction of the listener is in favor of a tone that is subject to a slight modulation effect; it seems to be more rich and satisfying. Undoubtedly there are subtle, and somewhat intangible, elements that enter into the determination of tone quality; and vibrato appears to fall in that category.

QUESTIONS

1. What determines the quality of a given musical sound?
2. Are the terms *harmonics* and *overtones* synonymous?
3. What is meant by *inharmonic partials?*
4. Why, in general, are inharmonic partials undesirable?
5. Each of two singers may have a pleasing voice when singing alone, but when singing a duet the musical results may be unpleasant. Why?
6. Of what utility are sound spectra?
7. Outline a method by which a musical tone may be artificially created.
8. Why does the tone quality of musical sounds in a motion-picture theater depend upon where one sits?

[1] The reader is referred to an informative paper by Louis Cheslock on Violin Vibrato, published under the general title, *Research Studies in Music*, by the Peabody Institute, Baltimore, Maryland. The paper appeared in the April, 1931, issue.

9 *Musical Intervals and Temperament*

9–1. *Musical Intervals*

In the preceding chapters we have been studying the laws which govern the behavior of a single train of sound waves, such as the propagation of one musical tone through the air. We are now to consider the relation of musical tones to one another; and this brings us more definitely into the domain of music.

Comparatively few persons are capable of recognizing the true pitch of a single musical tone, but many individuals are able to tell what the **ratio** of the frequencies of two tones is. Furthermore, most persons can recognize the fact that certain tones, when sounded together, or immediately following one another, produce a pleasing effect, while other combinations give rise to a decidedly unpleasant reaction. Of all the 20,000 frequencies with which the acoustician deals, only a comparatively small number are actually used in music. Certain particular frequencies have been selected, through experience, and built into a system of pitch intervals and scales for use in the noble art of music. **By the term musical interval we mean the ratio between the frequencies of any two tones.** For instance, one musical sound may have a frequency of 262 vibrations per second and another a frequency of 528. The frequency in the second case is to the frequency in the first case as 2 is to 1. Or again, one might be dealing with one tone whose vibration number is 330 and another whose frequency is 660. The ratio of their frequencies would in this case also be 2 to 1. Thus we see that the interval is the same in the two cases; it is the **ratio** which is significant, in this connection, and not the absolute frequencies of the two tones involved. As another example we might again take the 330 and another vibration number whose value is 495. The ratio of the higher to the lower note in this case is 3:2. In other words, the "vibration fraction" is 3/2. This means that one sonorous body makes three complete excursions while the other makes two. And so we might go on forming an innumerable number of intervals. But the experience of the human race has taught that only certain notes when sounded **at the same time,** or in quick succession, produce agreeable auditory effects; the intervals between such pairs of tones have been computed and have been named, as shown in the following table:

Unison	1:1	Minor third	6:5
Octave	2:1	Major sixth	5:3
Fifth	3:2	Minor sixth	8:5
Fourth	4:3	Major seventh	15:8
Major third	5:4	Minor seventh	9:5

The term *third* in the above list refers to the interval between the basic note and the third note above, as for instance C and E. Corresponding statements could be made concerning the terms *fourth*, *fifth*, etc. Not all of the above pairs yield the same degree of smoothness when heard simultaneously. In the following section we shall endeavor to find the reason for the fact that certain intervals give rise to a sense of physical and aesthetic satisfaction, while others produce a disagreeable auditory reaction.

9–2. *Nature and Cause of Dissonance*

When two or more tones, evoked simultaneously, produce a rough auditory sensation, we say that the sounds involved are **dissonant;** when auditory roughness does not obtain, the sounds are classified as being **consonant.** Dissonance implies harshness; consonance connotes tonal smoothness.

It is often comparatively easy to define a phenomenon, but not so easy to arrive at an understanding of the **cause.** Why is it that the notes C and E when simultaneously sounded on the piano, for instance, produce a smooth musical effect, while the sounding of C and D has quite the opposite effect? The answer to the foregoing question was given by Helmholtz. It was his judgment that **dissonance is due to the disagreeable sensation produced by beats.** To quote Helmholtz: "When two musical tones are sounded at the same time, their united sound is generally disturbed by the beats of the upper partials, so that a greater or less part of the whole mass of sound is broken up into pulses of tone, and the joint effect is rough. This relation is called dissonance.

"But there are certain determinate ratios between frequency numbers, for which this rule suffers an exception, and either no beats at all are formed, or at least only such as have so little intensity that they produce no unpleasant disturbance of the united sound. These exceptional cases are called consonances."

Helmholtz contended that in the middle register the number of beats which gave rise to the maximum roughness is 33 per second and that they can be detected when the beat frequency is as high as 132 per second. When the beat frequency exceeds that value their effect on a musical sound is inappreciable. Helmholtz likened the auditory effects of beats to the effect produced when a flickering light falls upon the eye. When beats occur between the limits mentioned the auditory nerve assembly is irritated and fatigued.

Here we have the opinion of a great investigator who had spent some eight years studying this and related questions. The subsequent work of Koenig and other investigators serves but to confirm and extend his findings.

From our preliminary study of the subject of beats (Sec. 4–3), it will be evident that two pure tones will give rise to beats only when their frequencies are not identical in value. But if we examine the situation when two complex tones are simultaneously sounded, it will be seen that beats may occur, even though the fundamentals of the two tones have the same frequency. This is caused by the fact that the overtones of one of the tones may beat with the fundamental of the other, or the overtones of one may beat with the overtones of the other, or a combination of these circumstances may obtain.

Let us see how this works out in several cases. For comparison purposes let us examine the interval known as the octave. The case in point can be most simply presented by displaying the harmonics involved in terms of the usual musical notation (Fig. 9–1), using half notes to represent the fundamentals and quarter notes to represent the upper partials.

We will assume that the tones are sounded on any instrument capable of yielding at least six overtones (seven harmonics). Only

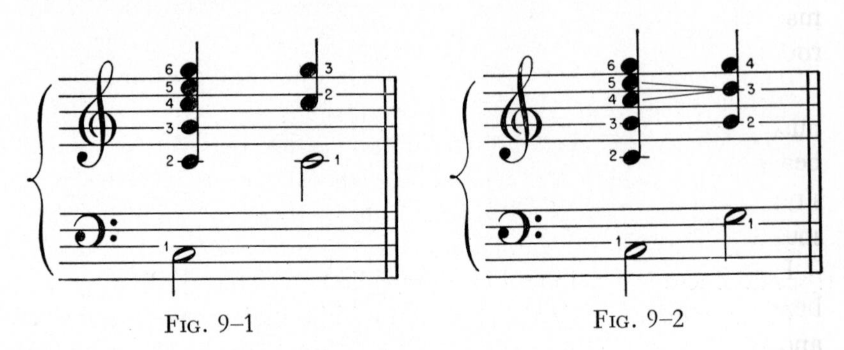

Fig. 9–1 Fig. 9–2

three harmonics belonging to the higher note are written down because the higher overtones of this note are not within beating distance of any overtones of the lower note. If the tuning is perfect each harmonic of the higher note corresponds to an upper partial of the lower note. It is thus evident that no dissonance can occur and that such an interval (the octave) is therefore perfectly consonant: the octave is unique in this respect. No interval narrower than an octave will yield absolutely perfect consonance.

Remembering the above statement, let us now examine the case of the fifth, an interval that comes as near to being a perfect interval as any of the several steps. Following the same plan as in the previous case, the musical diagram would appear as indicated in Fig. 9–2. In this case 3 and 2, and 6 and 4 coincide. But it will be noted that 3 of the higher note is within beating distance of both 4 and 5 of the lower note. If the fourth and fifth upper partials of the lower note happened to be weak the roughness would probably not be great, but in any event beats would probably exist, and hence this interval would not be as smooth as the octave. If we were to set up

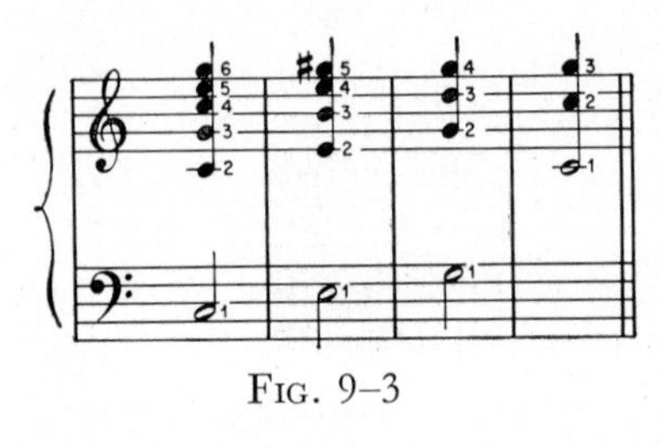

Fig. 9–3

a diagram of the fourth it would be found that this interval is somewhat less consonant than the fifth. In other words, these two steps, among others that might be cited, are imperfect consonances.

In this connection it will be profitable to consider the case in which four notes are involved, such as a major chord, like CEGC′.

An examination of the musical diagram appearing as Fig. 9–3 discloses the fact that the fourth harmonic of E is in unison with the fifth harmonic of C, but the third and the fifth harmonics of E are somewhat dissonant with the fourth and sixth of the tonic (C).

In the case of G, we see that there is only one partial, the third, which is not already present in the tonic C. This third will give rise to beats with the fourth and fifth of C, and to some extent detract from ideal consonance. The octave C′, it will be noted, introduces no partials which do not already exist in the tonic. While this major chord is, then, not perfect so far as consonance is concerned, it is comparatively free from disturbing beats; **this accounts for the satisfying harmony of the major chord.**

If space and time permitted, one could set up a corresponding chord which is known to be less harmonious than the above, and it would be seen that conditions would obtain giving rise to disturbing beats.

It may be therefore set down as an important fact that intervals and chords are consonant when beats are either absent or so faint as to be completely masked by the fundamentals of the tone involved. In this general connection it should be pointed out that **intervals and chords may be consonant in one part of the scale and yet be dissonant in another.** This is due to the fact that the number of beats per second which may develop will depend upon the position of the interval or chord in the scale. This is strikingly shown in the following instance. Take, for example, C_2 and E_2 having fundamentals of 65 and 82 respectively, and set down the frequencies of the fundamentals and the first two upper partials in each case, as per the following table.

	C_2	E_2
1st	65	82
2d	130	164
3d	195	246

It will be observed that there will be 17 beats per second between the fundamentals, which will probably not be disturbing. But when we come to the second harmonic in each case, we see that these partials will give rise to 34 beats per second, a frequency which will produce marked roughness. The third harmonic of C_2 will also develop a beat frequency of 31 with the first upper partial of E_2. Thus we see that an interval which was reasonably consonant in the middle register is quite unsatisfactory in a lower part of the scale.

9–3. *Dissonance in Special Cases*

Before leaving the subject of consonance and dissonance, it will be well to examine two or three special instances which are encountered in practice. The earliest forms of music consisted of a succession of single tones produced upon some simple instrument or by the singing voice. A tune sung by one voice or played on a single instrument, one note at a time, is what we know as a **melody;** ancient music was, and much oriental music still is, of this character. Now the major chord was known long before the Christian era when only single part melody was employed. Can one reconcile these two facts with the suggestion previously made to the effect that the major chord came into use because it was naturally consonant; that is, quite free from disturbing beats? The late Professor Sabine[1] has pointed out that harmony does enter into single part melody if and when the rendition takes place in a room, the reason being that a single sound persists for an appreciable length of time due to reverberations. This overlapping, then, of successive tones produces, to some extent, the effect of two or more notes being sounded simultaneously; hence, consonance and dissonance may figure in the case.

Another factor which may also enter into the situation is that there is a certain persistence of sound sensation. Therefore, even though there be no overlapping due to reverberations, tones sounded in rapid succession in a one-part melody might produce a dissonant effect.

In succeeding chapters it will be brought out that each instrument and each voice gives rise to a definite and characteristic sound spectrum—it has its own particular complement of overtones. Bear-

[1] See W. C. Sabine, "Collected Papers in Acoustics," p. 107.

ing this in mind, it is easy to see that a chord when played by two different instruments might be consonant, while if rendered by some other combination might be dissonant. The same might be said of two singing voices. This accounts for the fact that certain instruments and certain voices "blend," as we say, better than others. In other words, the whole question of consonance and dissonance involves the matter of the absence or the presence of beats. Furthermore, it should certainly be noted, the smoothest intervals are not always the most satisfying to the ear. This is evidenced by the frequent modern use of thirds and sixths in contradistinction to the use of the octave, the fifth, and the fourth. Indeed, actual dissonant intervals and chords are at times effectively made use of. Notable examples may be found in various contemporary compositions.

Our musical tastes apparently change with time. For instance, thirds and sixths were until comparatively recent times not used to any great extent by musicians. For some unknown reason the subminor seventh (4:7) has never been utilized to any great extent, although in some instances, at least, it is more consonant than the minor sixth (5:8). Possibly this and other intervals may come to form a part of musical composition. There is no virtue in mere newness or strangeness, but the acoustician and the musician should maintain an open mind with regard to the possible adoption of new forms of tonal expression.

9–4. *The Major Chord*

In the preceding section it was pointed out that certain pairs of notes, when sounded together, produce a pleasing auditory sensation. It is also a fact that certain groups of three notes will produce a similar effect; and one of the remarkable things about this circumstance is that the frequencies of such triads bear a definite but simple relation to one another, viz., that of 4:5:6. There are three such groups, as indicated below:

C:E:G = 4:5:6
G:B:2D = 4:5:6
F:A:2C = 4:5:6

From the above relations one may readily compute the vibration number of any tone, in terms of the basic note C, by means of the following simple equations:

$$\frac{E}{C} = \frac{5}{4} \quad \text{or} \quad E = \frac{5}{4}\,C$$

$$\frac{G}{C} = \frac{6}{4} \quad \text{or} \quad G = \frac{3}{2}\,C$$

$$\frac{F}{2C} = \frac{4}{6} \quad \text{or} \quad F = \frac{4}{3}\,C$$

$$\frac{A}{2C} = \frac{5}{6} \quad \text{or} \quad A = \frac{5}{3}\,C$$

$$\frac{B}{G} = \frac{5}{4} \quad \text{or} \quad B = \frac{15}{8}\,C$$

$$\frac{2D}{G} = \frac{6}{4} \quad \text{or} \quad D = \frac{9}{8}\,C$$

Any one of the above triads, taken together with the octave of the basic, or tonic, note, constitutes what is known as a **major chord;** thus, C, E, G, and 2C would be one major chord.

9–5. *Diatonic Scale*

The major diatonic scale is composed of three sets of major triads, thus:

C	D	E	F	G	A	B	C′	D′
4	. . .	5	. . .	6				
			4	. . .	5	. . .	6	
				4	. . .	5	. . .	6

The initial note, in this case C, is called the **tonic.** It is important to consider the relation of the vibration frequency of each note in the above scale to the tonic. If we assume that the tonic has a frequency represented by n, the successive notes will have the frequencies indicated in the following chart, thus:

Name of note	C	D	E	F	G	A	B	C
Frequency in terms of tonic	(1/1)n	(9/8)n	(5/4)n	(4/3)n	(3/2)n	(5/3)n	(15/8)n	2n
Ratio between successive tones (intervals)	9/8	10/9	16/15	9/8	10/9	9/8	16/15	

The vibration fractions (intervals) between successive notes (third line in above chart) are obtained by taking the ratio between two adjacent frequency values. For instance, the ratio of (9/8)n to n is 9/8; of (5/4)n to (9/8)n is 10/9, etc. An examination of the above table discloses the fact that, in the diatonic scale, there are only three intervals, viz., 9/8, 10/9, and 16/15. The first two intervals are referred to as **whole tones,**[1] and the last as a **half tone.** They are also designated as a **major tone,** a **minor tone,** and a **major** or **diatonic semitone,** in the order given. Strictly speaking, the last interval (semitone) is slightly greater than half of a major tone.

Reduced to mixed numbers the vibration ratios 9/8, 10/9, and 16/15 become 1.125, 1.111, and 1.067 respectively. When making computations this numerical form of the ratios will be found to be more convenient than the fractional form. Now in adding intervals, one multiplies the ratio values involved. By way of illustration, the interval between C and E is obtained by multiplying 1.125 (the interval C to D) by 1.111 (the interval D to E) which gives 1.250. If one proceeds in a similar manner a chart may be set up showing the intervals between C and the other notes of the scale as given in the first line of figures in the following arrangement.

	C	D	E	F	G	A	B	C′
Diatonic scale	1.000	1.125	1.250	1.333	1.500	1.667	1.875	2.000
Tempered scale	1.000	1.222	1.260	1.335	1.498	1.682	1.888	2.000

[1] The term *tone* means, in this connection, a **step,** and is approximately equal to one-sixth of an octave.

The diatonic scale is of very ancient origin, but this scale as now used was introduced by Zarlino, the first account of it being found in his "Istituzioni Armoniche," published in 1558. The diatonic scale appeals to the occidental as constituting a satisfactory basis for musical composition. This scale is by no means the oldest arrangement of musical intervals. The modern diatonic scale bears some resemblance to the so-called Pythagorean scale,[1] and probably is related to it, historically. The Greek scale is found to be quite suitable for melody—a succession of single notes—while the modern diatonic scale is better adapted for harmonic composition.

Each of the older races, which make a pretense to anything of a cultural nature, has adopted some system of note relationship. For instance, the Chinese divide the note cycle into twelve equal steps, corresponding to our semitones; but in practice they frequently use only five notes, corresponding to the black notes on our piano keyboard; that is, a pentatonic scale. One also finds the pentatonic division of the octave in some examples of Scotch music.

The Arabs divide their cycle into sixteen unequal intervals with the result that their music is entirely different from anything one finds among Western peoples. The octave and the fifth are utilized, and they frequently use quarter tones.

The Hindus divide their cycle into twenty-two steps, though they actually use only seven intervals. The octave, the fifth, and the fourth form a part of the Hindu system.

The Persians, who undoubtedly exerted an influence on Greek music, and therefore on our scale, employed very small intervals. Their octave was divided into twenty-four steps, which means that they must have used quarter tones.

Even so brief a review of musical history as the foregoing discloses the fact that about the only intervals common to the oriental and

[1] Pythagoras, who was the founder of theoretical music, used but two intervals in developing the scale which bears his name, viz., the tone and the semitone (hemitone). His tonal arrangement was

C		D		E		F		G		A		B		C
	9/8		9/8		256/243		9/8		9/8		9/8		256/243	

occidental scales are the octave and the fifth. It is therefore evident that there is no such a thing as a "natural" scale. This is not to be taken to mean that none of the intervals have a natural basis for their use. Probably the selection of the octave rests, to some extent at least, on the fact that a woman's voice is about an octave above that of a man's. Then again the octave is, in a way, a repetition of the fundamental, in the case of complex tones, and is often the most easily recognizable as an element in such a sound. Similarly, the third partial (second overtone) is just a fifth above the second partial. Frequently this tone can also be easily recognized when listening to a complex tone. While the fourth might also be said to have a somewhat similar natural basis for its more or less unconscious selection as a step in our scale, the remaining intervals have been evolved out of our experience and as a result of our temperamental tastes. We shall look into the question further in the chapter which follows.

9–6. *The Minor Chord and Scale*

In addition to the major triad and the major chord there is a so-called minor triad and a corresponding minor chord. The notes in each minor triad bear the relation to one another of 10:12:15. One can build up a series of vibration ratios based on a tonic as in the case of the major grouping (Sec. 9–4). A minor scale can also be worked out from these fractions, as set forth in the table below:

C	D	E	F	G	A	B	2C
$(1/1)n$	$(9/8)n$	$(6/5)n$	$(4/3)n$	$(3/2)n$	$(8/5)n$	$(9/5)n$	2
9/8	16/15	10/9	9/8	16/15	9/8	10/9	

A comparison of the major and minor scales discloses the fact that the same three intervals occur in both, but that the order of occurrence is somewhat different; the second and third are interchanged as are also the fifth and seventh. The vibration numbers for the two scales, using A = 440, are:

	C	D	E	F	G	A	B	C′
Major	264	297	330	352	396	440	495	528
Minor	264	297	316.8*	352	396	422.4*	475.4*	528

It will be noted that three frequency numbers (starred) of the minor scale differ from the corresponding values in the major scale; they are in each case **lower** in frequency. This should be kept in mind in reading the next section.

9–7. *The Tempered Scale*

Owing to the natural limitations of the voice of a soloist it may be necessary for the performer to use some tone higher than C as a tonic.[1] In such an event it is interesting and significant to compare the scale built on C **as a tonic** with one using, say, D as a fundamental. The following chart gives such a comparison:

	C	D	E	F	G	A	B	C′	D′
Key of C	264	297	330	350	396	440	495	528	
Key of D	. . .	297	334*	371*	396	445*	495	557*	594

It will be noted that four of the frequency values (starred) in the D scale do not agree with the corresponding numbers in the lower scale. It is thus evident that at least three additional tones would be needed if one were to employ the scale of D; that is, use D as a tonic.

It is entirely possible that a vocalist might use any one of the other notes as a tonic; and if we were to set up such a series of scales and compare them with the simple diatonic scale, we should find that several other interpolations would be necessary. Our study of the minor scale (Sec. 9–6) showed that additional tones would also be needed in order to provide for a system of scales based on minor chords. It therefore becomes obvious that, in order to provide for all possible changes of key, it would be necessary to intro-

[1] The making of such a change is known as "transposition."

duce a substantial number of notes into each octave—a number which would be entirely impracticable when dealing with instruments having fixed tones, such as the piano and most of the wind instruments; at least 36 notes to the octave would be required.

In order to avoid this practical difficulty, musicians have had recourse to a compromise procedure; they have adopted what is known as the **equally tempered scale.** The system provides for an octave having twelve **equal** steps or intervals. To accomplish this, five new tones are added to the original diatonic scale, as diagrammed in Fig. 9–4. These five tones serve as the *sharps* of the tones just below, and as the *flats* of those immediately above. In this system the common ratio of one frequency to the next is the twelfth root of 2, or 1.05946.

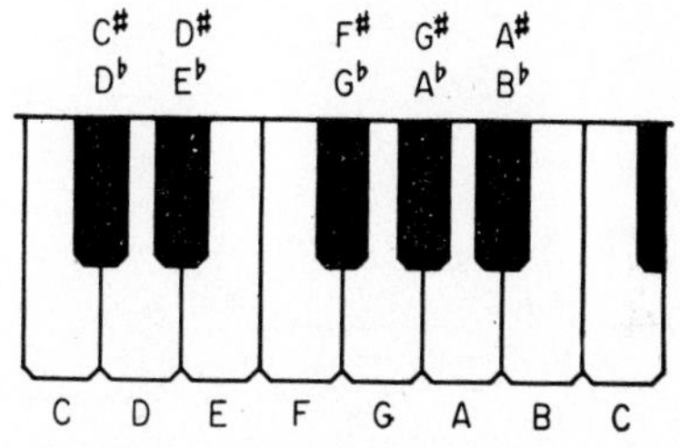

FIG. 9–4. Diagram showing a tempered octave.

But why the twelfth root of 2? Remembering what was said in the previous section to the effect that intervals are added by the procedure of multiplying their respective ratio values, it will be seen that if there are to be 12 equal steps in the scale the sum of these steps will be n^{12}, where n is the value of each interval. Further, this sum must be equal to 2 because C′ = 2C. That is n^{12} must equal 2 or, algebraically expressed,

$$n^{12} = 2$$

or

$$n = \sqrt[12]{2} = 1.05946$$

It will be recalled that the semitone interval in the natural scale has the value 1.067, as compared to the above value. In making computations the above numerical value of the semitone interval on the tempered scale is often taken as 1.06. By using the numerical value of the semitone of the tempered scale, the interval between C and the successive notes on the equally tempered scale can be readily computed. The ratios are shown in the last line of the chart on page 131. The intervals of the five added steps with respect to the tonic will have the following values:

C♯ or D♭ $1.000 \times 1.059 = 1.059$
D♯ or E♭ $1.122 \times 1.059 = 1.189$
F♯ or G♭ $1.325 \times 1.059 = 1.414$
G♯ or A♭ $1.498 \times 1.059 = 1.587$
A♯ or B♭ $1.682 \times 1.059 = 1.782$

One may now arrange a complete table showing the frequency ratios in the scale of equal temperament based on C as a tonic. In the following table *n* signifies frequency of C.

Note			Frequency ratio	Cents from tonic
	C		1.000*n*	0
C♯		D♭	1.059*n*	100
	D		1.122*n*	200
D♯		E♭	1.189*n*	300
	E		1.260*n*	400
	F		1.335*n*	500
F♯		G♭	1.414*n*	600
	G		1.498*n*	700
G♯		A♭	1.587*n*	800
	A		1.682*n*	900
A♯		E♭	1.782*n*	1000
	B		1.888*n*	1100
	C		2.000*n*	1200

Thus we have available a complete set of interval values by which one can determine the frequency of any note on any scale.

Any instrument tuned on the basis of the equally tempered scale has only one true interval, viz., the octave; all other intervals are slightly false. The thirds and sixths are somewhat sharp, and the fifths are slightly flat, as may be seen from an examination of the chart on page 131. From the mathematical nature of the case a 12-step octave is imperfect. Music played when employing such a system is necessarily somewhat inferior to that rendered in true intonation. However, the lack of exactness, with its resulting slight roughness, is tolerated for the practical reasons cited above. Unaccompanied playing on stringed instruments and on the trombone, as well as a cappella singing, can be done in just intonation, though even in such cases the performers sometimes depart from both just intonation and equal temperament. There are those who feel very strongly that music rendered through the medium of the equally tempered scale is decidedly inferior in quality to that played in true intonation. Professor Jones in his book "Sound," pages 70–74, quotes the opinion of several well-known musical authorities on the question. These excerpts are worth perusal. There are a number of practical difficulties connected with the construction and playing of a keyboard instrument that would render music in just intonation. However some progress has been made in solving this problem. In this connection the interested reader is referred to two papers by Professor Charles Williamson to be found in The Journal of the Acoustical Society of America, vols. 11 and 15.

The introduction of the equally tempered scale is usually credited to Johann Sebastian Bach, but it appears that this system of temperament was known long before Bach's time.

Before the equal temperament plan was introduced, there existed another system of temperament, introduced by Zarlino and Salinas, and known as unequal or meantone temperament. Under this system the more common scales were rendered fairly accurately and the remainder were disregarded. A performer was accordingly limited to certain keys only. At one time this system of temperament was extensively used, particularly by organists. For additional details on the history of both meantone and equal temperament the reader is referred to "Science and Music," Chap. 5, by Sir James Jeans, to the Ellis translation of Helmholtz, pp. 546ff., and to a scholarly historical treatise entitled "Tuning and Temperament," by J. Murray Barbour. This last work carries an extensive bibliography on this subject.

9–8. *Tuning in Equal Temperament*

As pointed out in Sec. 7–7, the tuning of such instruments as the piano can be accurately and rapidly accomplished by a visual method that utilizes the stroboscopic principle. Though the visual method is being increasingly used, many tuners still make use of the beat method.

In following the more conventional procedure (the beat method) the first step involves what is known as the "laying of the temperament." This consists in selecting an octave near the middle of the scale, say the one beginning with A_4, and tuning each string in that octave by means of beats between selected upper partials (overtones) of certain pairs of notes. The intervals most commonly utilized are fifths and fourths. If we keep in mind that the upper partials of a piano string, for instance, have frequencies that are about two, three, four, five, etc., times that of the fundamental, it will be easy to understand how the temperament of a piano may be set.

A preliminary step in the process of laying the temperament consists in damping two of the strings of each note by means of rubber wedges; thus only one string sounds when a key is struck. If we assume that the A_4 string has been adjusted to 440 against a standard fork, its third upper partial will have a frequency of 1320. The fundamental of E_5 (a fifth above A_4) will be 659, and its second overtone will have a frequency of 1318. Hence, if the interval from A_4 to E_5 is not close to a just fifth, there will be more than two beats per second, and the tension of the E_5 string must be adjusted until the number of beats is reduced to an acceptable value. If the interval from A_4 to D_5 is not close to a true fourth, there will be beats between the fourth overtone of A_4 and the third upper partial of D_5. Proceeding in a similar manner one may, if necessary, adjust the tension of the strings in the selected octave, and thus lay the temperament of the instrument as a whole. Each tuner has his own particular method of procedure; some go upward by fifths and downward by fourths, while others reverse this procedure.

Once the single string for each key has been properly adjusted the damping is removed and each of the several strings of a given note is adjusted to give zero beats with the string originally tuned.

(In research on the piano the author has found that the three strings of a given note will often depart from strict unison within a matter of a few hours after being tuned.)

Once the temperament is set, the strings of the remaining octaves are adjusted by octaves and unisons until they show no beats. However, in carrying out this procedure it becomes necessary to make certain compromises in the tuning process. This is because the actual strings possess a certain amount of physical stiffness. It therefore follows that the overtones of a struck string are not exact harmonics of the fundamental—there is a certain amount of what is called *inharmonicity* of the partials. The upper partials of a given fundamental are found to be slightly higher in pitch than the theoretical harmonics would be. Therefore in order to bring the upper partials of a given note into reasonably close consonance with the fundamental an octave higher a procedure known as "stretching" is carried out—the lower octaves are tuned slightly flat and the upper ones slightly sharp.

The procedure outlined above is followed to a greater or less extent in the tuning of an organ. For a detailed account of tuning practice the reader is referred to a work entitled "Piano Tuning and Allied Arts" by William B. White.

9–9. *Key Characteristics*

In closing our discussion of intervals and temperament it may not be out of place to mention the debatable subject of key characteristics. There are those who contend, for instance, that a composition written in the key of C major will have different tonal characteristics from the same composition if transposed to the adjacent key of D—that the one may convey to the listener a feeling of decisiveness or powerful resolve, while the other seems to suggest fullness or sonority. Some musicians claim to sense quite clearly such different characteristics and to ascribe certain emotional qualities to certain keys. Is there any objective basis for such an opinion? In attempting to answer this question there is one fact to be kept clearly in mind, and that is that on any instrument tuned in equal temperament the intervals (semitones) are all equal. As a result of this, the scales that represent the different keys differ only in pitch—the vibration ratios are identical. It therefore seems probable that, so far as pianoforte music is concerned, the special characteristics ascribed to individual keys are largely, if not wholly, subjective—a psychological phenomenon. However, in the case of the stringed instruments, such as the violin, it is conceivable that the different keys may possess definite characteristics. This could be due to the fact, as we shall see later, that each open string of a violin, for instance, has its own tone spectrum, and in playing in certain keys, C major for example, use is made of open strings more often than in executing a passage written in such a key as G♭.

QUESTIONS

1. What is meant by the terms *consonance* and *dissonance?*

2. What is the cause of dissonance?

3. What is meant by musical intervals?

4. Explain why there is only one musical interval in which consonance is perfect.

5. Name the most common musical intervals, and give their respective frequency ratios.

6. Given the frequency ratios of two intervals, how does one find the frequency ratio which is the sum of these two intervals?

7. What are the frequency ratios of the notes constituting the diatonic scale?

8. If G is 392 vps, what are the frequencies, respectively, of a minor third below and a perfect fifth above?

9. Why is a tempered scale necessary?

10. What is the frequency ratio between the successive notes on the tempered scale?

11. How many beats per second will there be between the seventh harmonic of C_3 and the nearest note on the tempered scale?

12. Briefly outline the procedure by which a piano is tuned.

10 *Musical Strings*

10–1. *Laws of Vibrating Strings*

Having considered, in the preceding chapters, the basic facts and laws connected with the generation and propagation of musical sounds, we are now prepared to pass on to a study of those generators of sound waves which are most widely used as sources of musical sounds. We shall consider in order: strings, air columns, rod, plates, and membranes.

The first generator to be examined is the string. A string, from the standpoint of acoustics, might be defined as a perfectly flexible filament of perfectly elastic solid material having a uniform cross-sectional area and stretched between two fixed points. It will at once be recognized that no physical string would be perfectly elastic and perfectly flexible, though it might be of uniform diameter. However, there are various materials such as steel and animal tissue which, when utilized as strings, give a close approximation to the ideal specifications.

The use of strings as a source of musical sounds dates back to the dawn of history. Indeed, there are many interesting and picturesque legends. Records have been found along the Nile showing a very early use of a simple form of the harp; in later times we read of the psaltery of the Israelites; the lyre was used by the Greeks, who ascribed its invention to Apollo; and Egyptian tradition ascribes the introduction of a similar instrument to Mercury.

However, it was not until Pythagoras, in the sixth century B.C., began his study of musical intervals that any systematic study of strings as generators of musical sounds was attempted. The invention of the monochord (Fig. 10–1 shows a modern form of this device) is commonly ascribed to him. With it he discerned the fact that if one part of a sonorous string is twice as long as another part, the shorter section will yield the octave of the longer part. He also discovered that the simpler the ratio of the two parts into which the vibrating string is divided, the more nearly do the resulting sounds approach perfect consonance. These discoveries served as the basis of much philosophic speculation concerning the nature and cause of harmony, but led to no search for quantitative laws.

Some two thousand years passed after the time of the famous Greek philosopher and mathematician before any real advance took place in the field of musical acoustics. Then in the seventeenth century, the Franciscan friar, Père Mersenne, worked out the laws

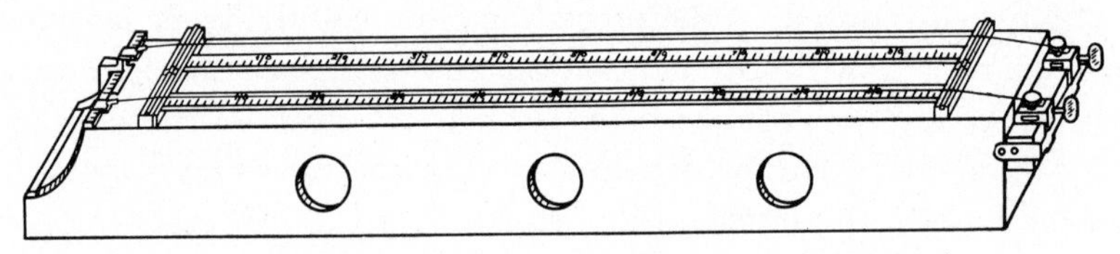

Fig. 10–1. Sonometer. (Courtesy of C. J. Ulrich.)

which relate the pitch of the sound produced by a vibrating string to its physical constants (Chap. 7). In his famous treatise "Harmonie universelle," Mersenne set forth four important laws of vibrating strings, which are as follows:

1. The number of vibrations per second is inversely proportional to the length of the string.
2. The number of vibrations per second is proportional to the square root of the tension to which the string is subjected.
3. The number of vibrations varies inversely as the thickness of the string.
4. The number of vibrations is inversely proportional to the square root of its density.

Whenever a string vibrates there will of course be a node at either end. When it is caused to vibrate as a whole there will be an anti-node, or loop, in the central region, as indicated by A in Fig. 10–2; a standing wave is developed, and hence the length of the string is equal to half a wavelength. If excited at the proper points, as we shall see presently, it may also break up into two or more vibrating segments, as seen in B, C, and D of the diagram, the length of each loop being equal to a half wavelength of the sound emitted by that segment.

One may, then, summarize the situation by means of a general expression

$$L = n\frac{\lambda}{2}$$

where λ is the wavelength associated with each segmental pattern, L the length of the string, and n any whole number. We have previously seen (Sec. 2–1) that

$$f = \frac{s}{\lambda}$$

where f is the frequency, s the speed of propagation, and λ the wavelength. By combining the last two expressions we get

$$f = \frac{n}{2L}s$$

It is known that the speed of a transverse wave disturbance in a stretched string is given by the relation

$$s = \sqrt{\frac{T}{m}}$$

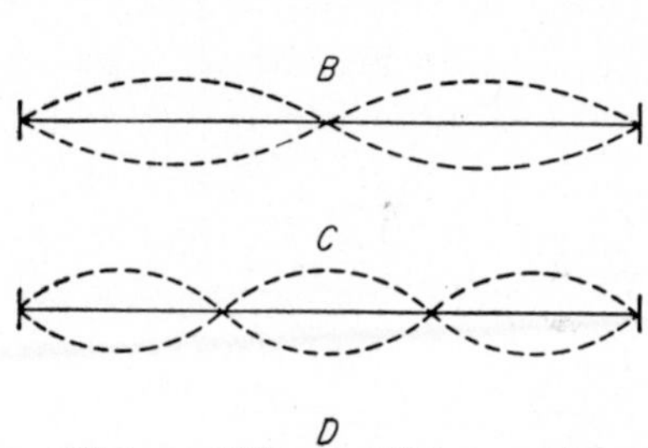

Fig. 10–2

where s is the speed, in centimeters per second; T the linear tension, in dynes, to which the string is being subjected; and m the mass of the string per unit length, in grams per centimeter. If now we combine the last two equations we arrive at the expression

$$f = \frac{n}{2L}\sqrt{\frac{T}{m}}$$

which gives the frequency, in vibrations per second, of any segmental vibration of a string, and hence the frequency of the resulting sound wave. Referring again to the verbal statement of Mersenne's laws governing the vibration of strings, it will be noted that the above expression constitutes a mathematical summary of those highly important relations. Knowing the magnitude of the tension, and the other factors involved, one can readily determine the frequency of the fundamental and any other harmonic that may appear. For the first mode of vibration (fundamental) $n = 1$; for the second mode of vibration (first overtone, octave) $n = 2$, etc., for successive cases, as shown in Fig. 10–2.

At this point in our discussion an important question presents itself: What determines the manner in which a string will develop segmental vibration? From the mechanics of the case there will always be a node at each end of the string. But it is evident that a node cannot exist at the point where the string is excited—that point will obviously be an antinode or loop. In the simplest case (Fig. 10–2*A*) we may assume that the string is being excited at its mid-point. If the string were to be plucked, for instance, at a point that is one-fourth of its length from one end, we would have the condition shown in Fig. 10–2*B*, a node appearing at the mid-point. If the string be plucked at a point one-sixth from one end, the situation depicted in Fig. 10–2*C* would obtain, giving rise to two nodes in the body of the string. If the excitation takes place at a point one-eighth from one end there will be three nodes, as seen in *D*. It will thus be evident that the median point is a node for all even-numbered harmonics, and a loop for all odd-numbered harmonics. To put the case another way, if we excite a string at its mid-point, all even numbered harmonics will be absent—only odd harmonics will be present. Likewise, if the string is excited at a point one-fourth of its length from one end, the second harmonic (octave) will be present in strength, but the fourth will not exist, and the third will appear only faintly.

One could greatly extend such an analysis, but the examples cited will suffice to show that the point at which excitation takes place will be an important factor in the determination of what overtones are present when a string is caused to vibrate, and hence will constitute a factor in the determination of the timbre of the resulting tone. The effect that the point of excitation has on the resulting waveform is illustrated by the oscillograph records in Fig. 10–3.

Mersenne's laws find practical application in connection with the functioning of all stringed instruments. We have seen that, other factors being held constant, the pitch of the tone evoked is a function of the length of the string. On instruments such as the guitar, the mandolin, and the banjo, frets are provided whereby the player is enabled to alter the length of the string by definite and fixed amounts; while in the case of the violin, cello, etc., the performer has complete control of the length of the string by pressing it against the fingerboard at any point desired. Hence, the great

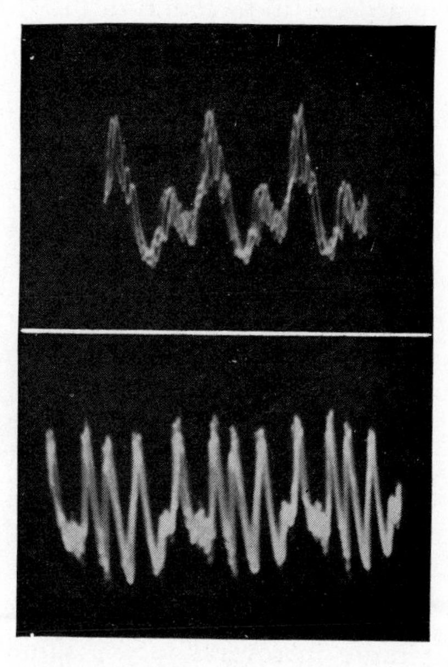

FIG. 10–3. Records showing the effect of point of excitation of a string on the waveform. *Upper trace*, sonometer string when plucked at mid-point; *lower*, at one-eighth the distance from one end.

flexibility of such instruments, and the possibility of playing them in true intonation.

In order to produce low tones without employing unduly long strings, certain ones are made of greater cross section, and thus of greater mass per unit length, the added mass in some instances being attained by wrapping a string with closely wound wire as in the bass strings of the piano. In the instruments of the violin family the strings of a given instrument are all of equal length; while in the case of the piano the length varies, as well as the cross section and tension. The tension on a single piano string is from 100 to 500 lb, and the total force which the sounding board must sustain is of the order of 25 tons.

It is a common experience to have the pitch of a string change as the temperature changes. This is owing to the fact that an increase in temperature will cause the string to expand and hence lessen the tension; the pitch, therefore, goes down as the temperature goes up.

At least in the case of a piano, relative humidity affects, to some extent at least, the tuning of the instrument. Dr. R. W. Young has given this subject extended study and finds that, in the case of the instrument investigated, the pitch rose 0.3 per cent for each increase of 10 per cent in relative humidity. Dr. Young's findings are given in a paper to be found in the November, 1949, issue of the *Journal of the Acoustical Society of America*, p. 577.

10–2. *Complex Vibrations in Strings*

Thus far we have considered but one mode of vibration in strings; the string, we have assumed, vibrated as a whole giving its fundamental tone or it vibrated in some definite number of segments, thus yielding some particular tone of higher pitch than the

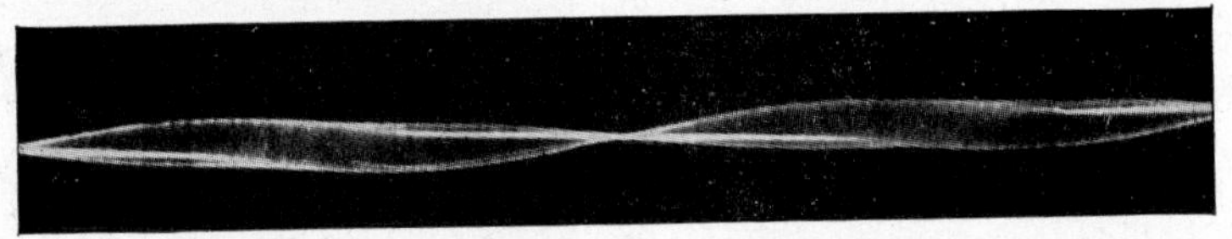

FIG. 10–4. Photograph of a string vibrating simultaneously in several modes. (Reproduced by permission from "The Science of Musical Sounds" by D. C. Miller, published by The Macmillan Company.)

fundamental. The question which next presents itself is: Can a string vibrate in several modes simultaneously; that is, can it vibrate at its fundamental frequency and at the same time produce segmental vibrations? The answer is that it can; and the results of such complex vibrations are highly important, as we shall see.

By suitably exciting a string it is possible to develop standing waves of a complex character, indicating that the string is not only vibrating as a whole, and thus yielding its fundamental tone, but that segmental vibrations are taking place at the same time. Figure 10–4 is a photograph of a string vibrating simultaneously in several modes. If a resonator having a broad response were to be associated with a string vibrating in such a manner, we would hear a complex tone, consisting of a fundamental accompanied by various upper partials.

The order of the partials which are evoked, and their relative intensity, depends upon the character of the excitation; upon the point at which the string is excited; and upon the density and elasticity of the string involved. In the piano, for instance, the hammer is arranged to make contact with the string at a point that is between one seventh to one ninth of the effective length of the string from one end, the purpose being to eliminate or greatly weaken

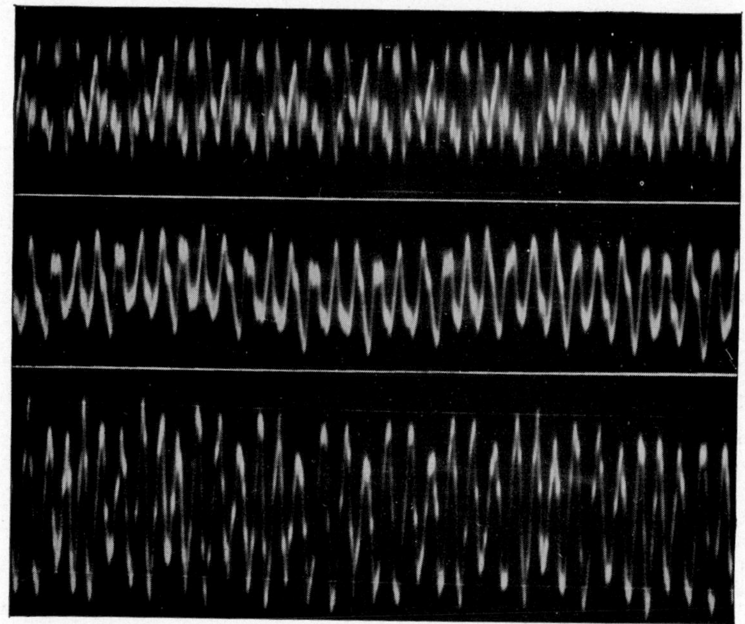

FIG. 10–5. Waveform of the sound emitted by a sonometer string when bowed (*upper record*), struck, and plucked; frequency 131 cps.

all partials above the sixth, particularly the seventh and ninth, which are more or less dissonant.

A given string will yield a tone of decidedly different quality depending upon whether it is struck, bowed, or plucked. This is strikingly shown by the oscillograms appearing in Fig. 10–5. The series of upper partials present, and their relative intensity, will also depend upon the thickness of the string and upon the nature of the material composing the string. Thick strings, because of their rigidity, tend to inhibit the formation of high upper partials while thin strings yield a large number of high overtones.

The reader is doubtless familiar with the marked difference in the timbre of a tone emitted by a metallic and a catgut string. The high upper partials are more quickly quenched in the case of the gut string, due to its relatively poor elasticity.

10–3. *Longitudinal Vibrations in Strings*

In the preceding sections of this chapter we have been considering only the **transverse** vibrations of strings. Experience shows that it is also possible to establish **longitudinal** vibrations in such a sonorous body.

Suppose that we have a steel wire, as a string, on the sonometer. If we stroke it lengthwise by the use of a piece of chamois on which there is some powdered resin, it will be found that the string will yield a loud note of high pitch. If we shorten the active portion of the wire by 50 per cent and again excite it, the note elicited will be the octave of the one previously evoked. Proceeding in this manner to test still shorter lengths, we shall find that the law governing the relation between frequency and length is the same for both longitudinal and transverse vibration. Here, however, the similarity ceases. If one changes the tension, it is found that the pitch remains practically the same, except for extreme values of tension. If the several gut strings of a violin be stroked longitudinally by means of the bow, the resulting frequency is practically the same for all of the strings, thus showing that the mass of the string has little if any effect upon its frequency.

Because of the fact that a string in undergoing transverse vibrations necessarily changes in length, it must follow that a limited amount of longitudinal vibration must accompany the transverse motion. It is said that these longitudinal vibrations may occasionally be heard when the A string of the violoncello is sounded. It is not, however, these incidental longitudinal vibrations that are of serious moment, but rather those longitudinal vibrations which the unskilled player brings into being by inadvertently slipping the bow lengthwise on the string in the course of regular playing. The tones resulting from such longitudinal excitation are of high pitch and decidedly dissonant; hence the necessity of avoiding this undesirable type of vibration.

QUESTIONS

1. What determines the pitch of a string?

2. If two strings, one 100 cm long and the other 200 cm, are of the same material, cross section, and tension what would be the interval between the fifth harmonic of each?

3. What harmonic is absent when a string whose length is 32 cm is excited at 4 cm from one end?

4. A string 120 cm in length gives a pitch of C when bowed. Where should the frets be placed so that the diatonic scale can be played?

5. What two points on a vibrating string are nodes, regardless of the point at which the string is excited?

6. Why are the strings at the bass end of a piano made of heavier material than those yielding the highest notes?

7. Is the timbre of the sound emitted by a string the same when it is bowed as when it is plucked? Why?

11 *Stringed Instruments*

11–1. *The Violin*

The list of musical instruments which depend upon the string as a generator of sound is quite large, and may roughly be divided into three groups, the classification depending upon the manner of exciting the vibratile member. In one group we have those instruments in which the string is plucked, the most common examples being: the harp, the guitar, the mandolin, and the banjo. A second group would include those in which the string is struck, the most notable example being the piano. The third group includes those instruments which are made to sound by means of a bow, and represented by the violin, the viola, the violoncello, and the bass viol.

FIG. 11–1. Violin.

Since space does not permit an examination of all of these sources of musical sounds, we will confine our attention to certain representative members of each type.

First let us consider the violin. The exterior of this instrument is familiar to everyone; some of the more important features are, however, not so well known. The instrument, a representative model of which is shown in Fig. 11–1, is provided with four strings, three of which are of plain gut and the fourth wound with fine wire to give it sufficient mass.[1] The front, or "belly," is usually made of

[1] Some performers use two wound strings, one being wrapped with aluminum and the other with silver. Occasionally three gut and one steel strings are used, Certain violinists have successfully employed four steel strings.

spruce or pine, the back of maple or sycamore, and the bridge and ribs of maple. Attached to the underside of the top member, and extending about two-thirds of its length, is a small strip of wood called the "bass-bar." The bridge stands on two feet in a position mid-way between the two "*f*" holes in a region where the belly is most mobile. One foot of the bridge rests directly over the bass-bar. Beneath the bridge between the belly and back of the instrument is located the "sounding post," a short wooden rod which is in firm contact with the wood plates which form the two larger surfaces. See Fig. 11–2. There is some dispute as to the optimum position for this member of the structure. The most common position for the post is a little at the rear of the leg of the bridge on the E string side. Investigation has shown that the chief function of the sounding post is to transmit vibrations from the belly to the back of the instrument. It is necessary to adjust the position of the post for each particular instrument. The bridge serves to convey the vibrations of the strings to the body and thus in turn to the enclosed air; the bass-bar serves to strengthen that part of the belly to which it is fastened.

Fig. 11–2. Cross section of the body of a violin, showing the position of the sound post and bass-bar. (*The Philips Technical Review.*)

The absence of definite frets makes it possible for the player to control individually the pitch of each note, by pressing the string (called stopping) against the finger board with the fingers of the left hand. This arrangement also makes possible the gliding of a note continuously from one pitch to another—a procedure called **portamento.** It is possible for the player to produce high notes without utilizing extremely short portions of a string. This is accomplished by either lightly touching a string by a finger of the left hand at the mid-point of the otherwise "open" string or at a point one-third or one-fourth, etc., from the lower end; or by pressing a string firmly against the finger board with one finger and also lightly touching it with another finger at some point near the bridge. In the first instance what are called **natural harmonics** are evoked, while in the latter case the harmonics developed are referred to as **artificial harmonics.** (Why these names?)

The use of a resined bow of horsehair as an exciting agent makes possible the sustaining of a note for relatively long periods of time—a feature which is in sharp contrast to the corresponding aspects of execution with instruments in which the string is struck or plucked. The quality of tone evoked depends, to some extent, upon the position of the bow's contact with the string; upon the speed of the bow movement; and also on the pressure. Bow pressure is more effective in modifying tone structure than is bow speed. Increased bow pressure tends to increase the intensity of the higher partials. The point at which the bow is applied to the string varies considerably with different players, and, for a given performer, often changes from one passage to another. However, the most common position is about ½ in. from the bridge. If a string is bowed near the bridge, the higher harmonics are relatively prominent and the tone has a more "cutting" quality. If the point of bow contact is farther away from the bridge the higher upper partials sound less strongly and the resulting tone is more suited to the execution of pianissimo passages.

The amplitude of the string's vibration, and hence the intensity of the resulting sound, depends upon the amount of bow hairs which are in contact with the string, and upon the velocity of the bow; increased bow speed gives rise to an increase in sound intensity. The sound intensity is affected by the point of bow contact, the loudest tones being evoked when the string is bowed near the bridge. Bow pressure as such is, however, not an important factor in the control of intensity, especially in legato playing.

In discussing the subject of bowing it is interesting to consider the mechanical process by which the string is maintained in vibration. This process involves two related steps or stages.

As the bow is moved across the string it drags the string with it, because of friction, until the restoring force due to the elasticity of the string finally causes the string to slip backward toward, and overshoot, its original position. This two-stage process is repeated very rapidly as long as the bow is in contact with the string. The resulting vibratory motion of the string is known to have the waveform indicated in Fig. 11–3. The longer portions of the graph represent that part of the motion when the string is being moved

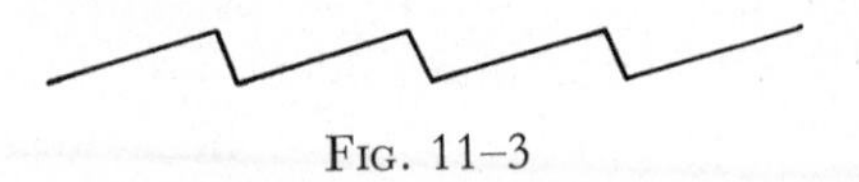

FIG. 11–3

because of the frictional force developed between the bow and the string. The shorter sections represent that part of the complete motion when the string is slipping. A vibratile member having such a saw-tooth type of motion will, from its nature, contain not only the fundamental mode but also a large number of both even- and odd-numbered upper partials. It is the function of the violin body, and the enclosed air, to act as a composite resonating system to select and to enhance the desirable harmonics thus developed and to suppress the concomitant undesirable components.

Research has disclosed that the vibratory motions of the bridge, of the belly, and of the enclosed air are not identical with those of the strings themselves. The vibrations of the belly appear to be largely responsible for the characteristic tonal quality of the violin.

The resonance of the body and enclosed air, while extremely broad, does show various resonance maxima, particularly one occurring at a frequency at or near 260 and another in the immediate region of 500. A rough line on the natural resonance point of a given instrument may be got by gently blowing across one of the *f* holes or by sounding a series of tuning forks near the same place. In some violins one encounters a peculiar case of sharp resonance. In these instances the body of the violin as a whole, or the confined air, or a combination of these elements, appears to be thrown into pronounced vibration at some particular frequency (usually not the fundamental of the wood) with the result that a strong and more or less uncontrollable tone is emitted from the instrument. This note is known as the "wolf note" from the howling effect which it produces. This phenomenon is most apt to be encountered in a poorly constructed violin, though it may occur to a lesser extent in instruments of the highest quality. Players naturally strive to avoid exciting the wolf note.

The waveforms characteristic of the four "open" strings of the violin are shown in Fig. 11–4. The records are of a high-grade instrument played by a skillful musician. It is to be noted that each string shows a distinctive waveform, and this implies that each string exhibits a definite individual timbre. It accordingly follows that each string will have its own sound spectrum. In Fig. 11–5 is shown a series of four spectra, set up from the accurate data obtained by Dr. Arnold Small, and originally disclosed by him in a

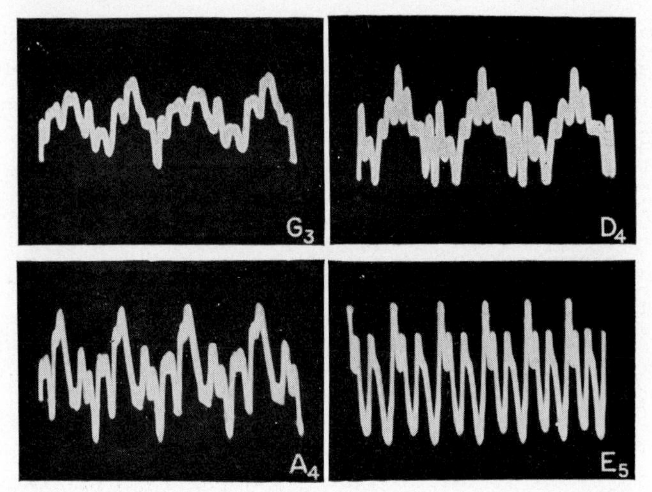

FIG. 11–4. Waveforms of the four open strings of a violin.

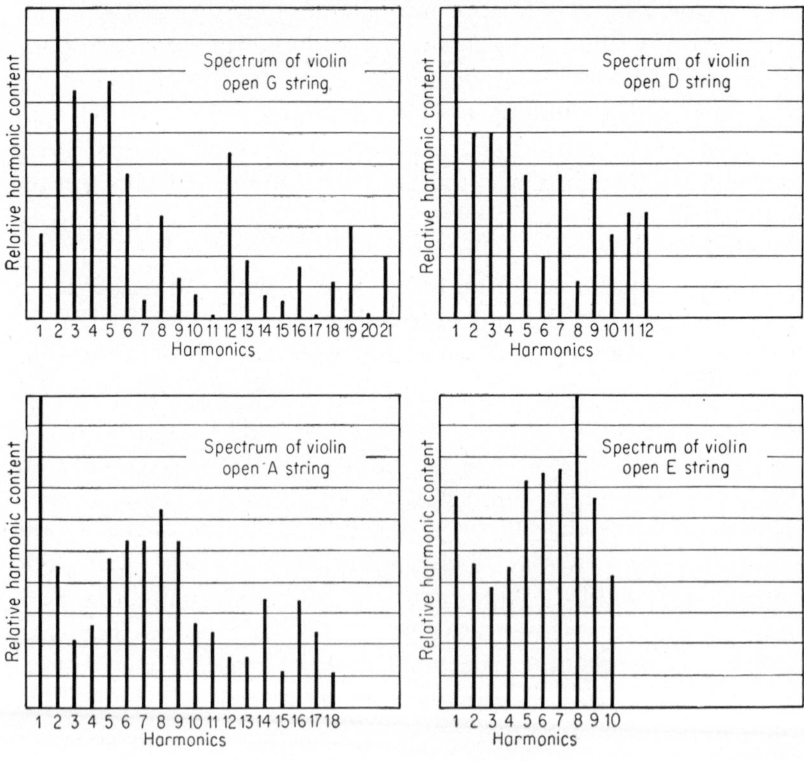

FIG. 11–5

paper appearing in the *Proceedings of the Music Teachers' National Association for* 1938. Dr. Small reports that "the violin was a Zanoli of about 1740, and was strung with steel E, gut A, aluminum-wound D, and a silver-wound G string." In securing these data all tones were played legato, mezzoforte, up-bow, and without vibrato. It will be observed that **the tonal color is entirely different for each string.** A study of these spectra discloses the fact that the normal violin tone is characterized by the presence of many upper partials, more than 20 in the case of the G string. In the case of two strings (E and G) the fundamental is not the loudest component. Because of the fact that each string has its own tone spectrum a particular passage is frequently played on a particular string in order to take advantage of its particular timbre.

The open strings of the violin yield the notes G_3, D_4, A_4 and E_5; and the normal pitch compass of the instrument lies between G_3 and B_7, a range of over four octaves. This range can be somewhat extended by evoking certain harmonics, as previously explained.

It is important that each string of a violin shall be as uniform in cross section as possible, otherwise the fingering would have to be altered slightly to compensate for the variation in thickness of the vibrating portion. Hence, when a string wears and eventually breaks, the player often replaces the entire set. Since all gut strings usually show a slight taper from one end to the other, it is important that all strings be installed with the taper in the same direction, otherwise the same stops on several strings would not be side by side.

The **mute** is a small notched clamp made of wood or metal which is affixed to the bridge and serves to "soften" the tone generated. The timbre and the over-all intensity are changed by the use of this device, as is shown by the records appearing in Fig. 11–6. The mute reduces the number and intensity of the higher partials and may strengthen the fundamental. Figure 11–7 should also be examined. The mute apparently accomplishes these ends by decreasing the amplitude of the vibrations of the bridge.

Certain tonal effects are produced by plucking the strings instead of bowing—**pizzicato** excitation it is called. The tone quality is decidedly different when using this method of causing the string to vibrate than when bowing is employed. Figure 11–8

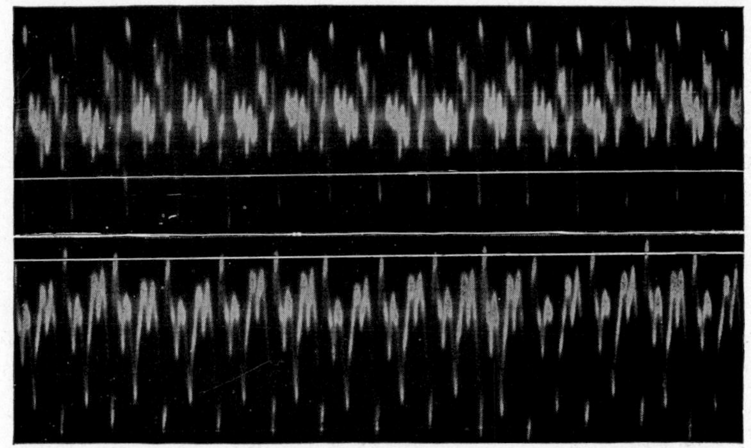

Fig. 11–6. Waveform of muted (*lower record*) and unmuted G string of a violin.

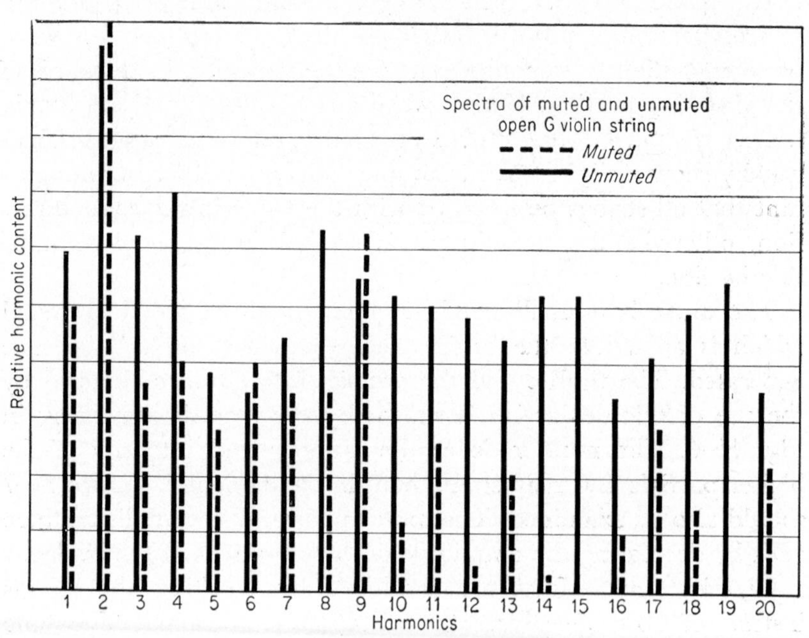

Fig. 11–7. Spectra of muted and unmuted open G string of a violin. (After Small.)

clearly shows the difference in waveform when the two methods of excitation are used. The tone quality of a plucked violin string depends to some extent on the point of excitation, as indicated by the waveform shown in Fig. 11–9. Violinists are often somewhat careless in playing pizzicato notes. Apparently some performers are under the impression that the point at which the string is plucked has no bearing on the resulting tone quality—an erroneous opinion, as the records show. The foregoing observation holds for

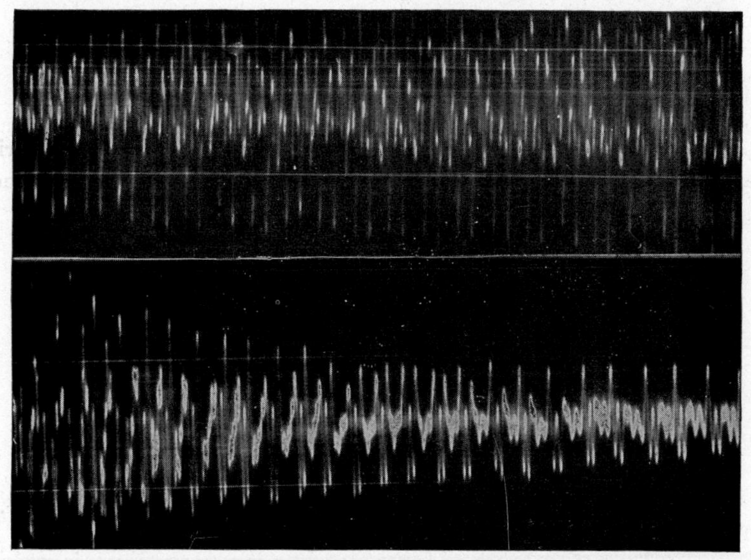

Fig. 11–8. Waveform of bowed (*upper record*) and plucked G violin string.

the players of all stringed instruments. It is to be hoped that a reform in this respect may be accomplished.

In rendering a passage it is sometimes necessary to reverse the bow during the execution of a given note. A photographic record of such a transition appears in Fig. 11–10. It will be noted that, as a result of the velocity of the bow becoming zero, for something like one two-hundredth of a second, the amplitude of the string's vibration is nearly zero. An examination of the record also shows that the waveform is turned completely over, showing that the phase undergoes a reversal, i.e., a change of 180°. However, the ear does not detect any change in quality because of this phase shift. Nominally the waveform, and hence the timbre, is the same in "up-bow" and "down-bow," but actually few violinists are capable of producing exactly the same waveform and intensity under the two circumstances mentioned. In Fig. 11–11 may be seen the waveforms of a good instrument that resulted from the

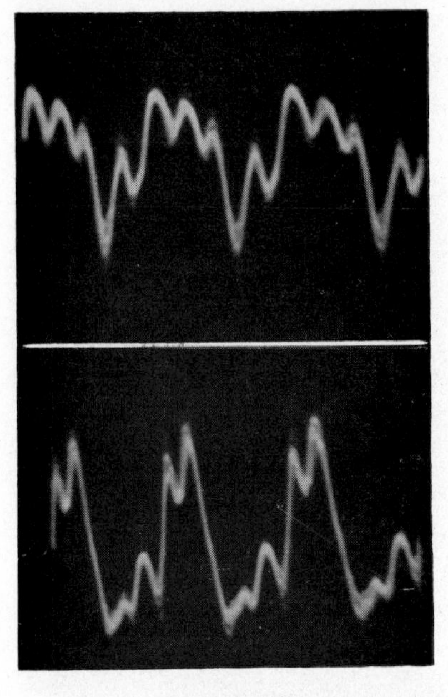

FIG. 11–9. Waveform of violin string when plucked at different distances from the bridge.

up-bow and down-bow movements when played by a skilled performer. It will be noted that the timbre is slightly different in the two cases. Obviously the matter of bowing technique is an important feature of violin playing. In this connection the reader

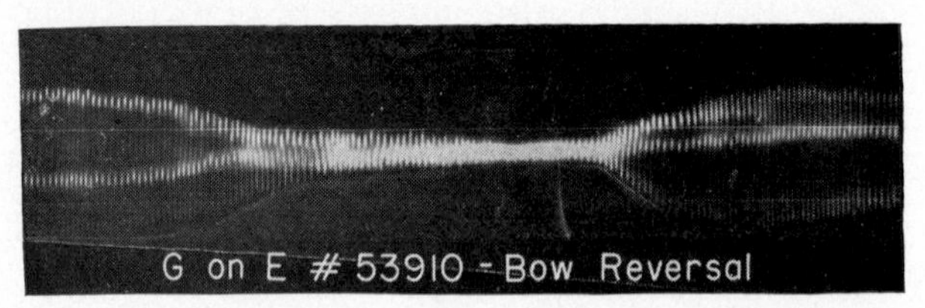

FIG. 11–10. Waveform of violin string when bow is reversed.

should consult the paper on the violin by Dr. Small, referred to on page 153.

The violin, as we know it today, appears to be the result of long generations of empirical development dating back to the earliest periods of oriental culture. The art of violin making reached its zenith during the seventeenth and eighteenth centuries. It was during that period that the famous Italian artisans produced their highly prized models. The names of Andrea, Antonio, and Nicolo Amati (three generations), Stradivari and Guarneri are among the names well known to lovers of highly prized instruments made by master craftsmen. After the death of Stradivari (1737) the art of violin making declined; most of the instruments made since have been copies of the original masterpieces. Occasionally, probably by chance, a fairly good model has been produced, but no one

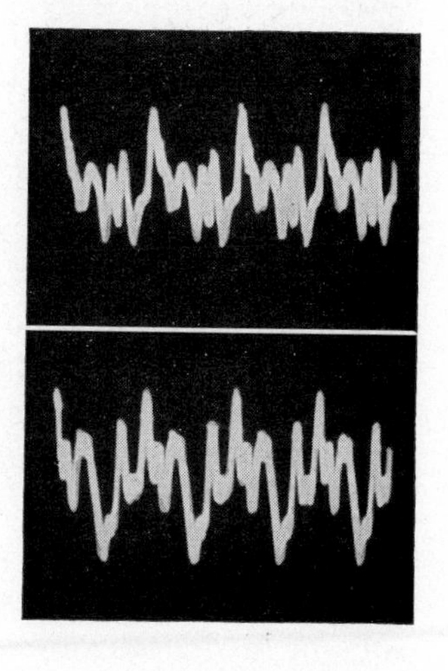

FIG. 11–11. Waveforms of violin string (A 440) when played up-bow (*top*), and down-bow (*lower*).

could tell why some were relatively good while others were of indifferent quality.

In recent years, owing to advances in research facilities, a number of serious attempts have been made to determine by scientific means the basis of these differences. For instance, it was formerly thought that the kind of varnish used had a bearing on the tone quality of stringed instruments, particularly in the case of the violin. But the results of recent scientific investigations indicate that the kind of varnish, as such, does not affect the end result.

It is, however, reported that by the use of X rays certain important facts have been brought to light about the grain of the wood used by the older craftsmen in the construction of the much-sought-after instruments.

H. Backhaus[1] in Germany has applied modern methods of sound analysis to the tones produced by a number of famous Italian violins; and more recently Poul Jarnak, of Copenhagen, himself a maker of high-grade stringed instruments, has, perhaps for the first time, made effective use of modern scientific methods of analysis in the design and construction of violins and cellos of good quality. His technical findings are set forth in a valuable paper published in the *Journal of the Franklin Institute* for March, 1938, to which the interested reader is referred. Prof. F. A. Saunders, of Harvard University, has also carried on extensive studies along these lines. A paper by Professor Saunders, which appeared in the January, 1946, issue of the *Journal of the Acoustical Society of America*, is of interest in the above connection.

In a more recent paper by Professor Saunders there is disclosed a very important discovery made by him in the course of his extensive studies on the violin. He found that by cutting a slight groove around the edge of the top (on the under side) of the instrument, the loudness of the instrument was increased about 2 db, and, at least in one case, the wolf tone was eliminated. Furthermore, as a result of this structural modification, the tone quality was materially improved, so much so in fact that the tone of an inexpensive violin was, as a result of such a mechanical change, made comparable to that of a high-grade instrument. Professor Saunders' original paper embodying this and other important results of his investigations on the violin will be found in the *Journal of the Acoustical Society of America*, vol. 25, pp. 491–498, May, 1953.

It appears probable that the investigations of the character above referred to will, in the near future, make it possible to produce, perhaps on a mass basis, violins and other stringed instruments whose tone quality and ease of playing will be equal to, and perhaps surpass, those of instruments made by the older master craftsmen.

We have dwelt upon the construction and playing technique of the violin at some length, first because the violin section is the basis of orchestral instrumentation; and second, because this instrument is inherently capable of producing an extremely wide variety of tone color; in this respect it surpasses all other instruments.

[1] H. Backhaus, *Z. Teck. Fiz.*, vol. 8, p. 509, 1927.

Indeed the violin constitutes, in effect, twelve instruments in one—each string has its own spectrum, as we have seen, and one may evoke a tone by any one of three methods. In the hands of a skilful musician it has almost unlimited capabilities of conveying to our senses the most profound and ennobling emotional concepts.

11–2. *Viola, Violoncello, Double Bass*

What has been said in the preceding section about the violin is, in the main, applicable to the viola and the violoncello. The **viola** might be said to be an alto violin. It is somewhat larger than the violin and has heavier strings, thus making it an instrument of lower range; it is tuned a fifth lower than the violin. Its four open strings are tuned to C_3, G_3, D_4, and A_4. Its range extends from C_3 to C_6, a span of three octaves, though its higher notes are seldom used. Its characteristic waveform is shown in Fig. 11–12; its spectrum is given in Fig. 11–13.

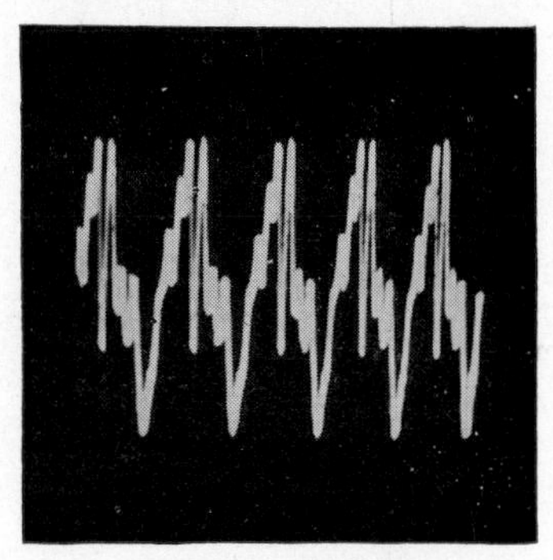

FIG. 11–12. Waveform of viola, open A string.

The viola is seldom employed as a solo instrument, but occasionally is used in the rendition of short passages with pleasing effect. The viola is frequently assigned important melodies in orchestral compositions. One fine example is to be found in a part written for it by Saint-Saëns in his "Reverie du Soir." Weber has given the viola an accompaniment to Annie's Recitative in Act II of *Der Freischütz*. Berlioz employs the viola in a solo part in "Harold in Italy."

The **violoncello** (Fig. 11–14) is the tenor of the strings. It is tuned an octave below the viola, thus making it an octave and a fifth below the violin. The four strings are tuned to C_2, G_2, D_3, and A_3. It has a range of more than three octaves, extending from C_2 to about E_5, but in skilled hands the upper limit may be somewhat extended. The strings of the cello are much thicker than those of the violin and viola, and about twice as long. Its timbre is such as to make it stand next to the violin as an instrument of

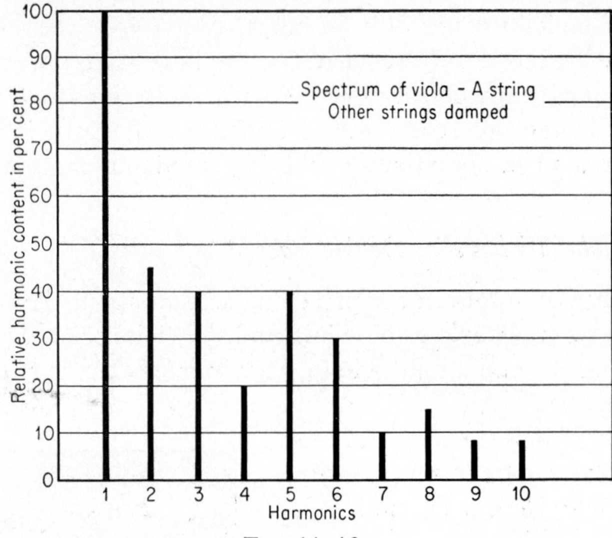

Fig. 11–13

great orchestral utility. Its tonal quality renders it particularly valuable for melodic expression; and it is often used effectively as a solo instrument. Figures 11–15 and 11–16 show the characteristic waveform and spectrum, respectively, of the A string. Each string has its own spectrum and corresponding color, as in the case of the violin. It will be noted that its sound spectrum is characterized by the presence of several strong overtones, the second of which is as strong as the fundamental. There was a time when the cello was equipped with five strings; in fact, Bach composed for such an instrument. Recently interest in such a type of violoncello has been revived. The addition of a fifth string facilitates execution on this instrument. Possibly, with the aid of scientific methods,

Fig. 11–14. Violoncello. (Carl Fischer Musical Instrument Co., Inc.)

a practical five-stringed tenor orchestral unit may be developed. The cello is widely used for both melodic and solo purposes. Perhaps the most popular solo part ever written for this instrument is "The Swan" in the orchestral fantasy, *Carnival of the Animals*, by Saint-Saëns. Grieg in "Anitra's Dance" of his *Peer Gynt Suite* gives the cello a beautiful independent melody.

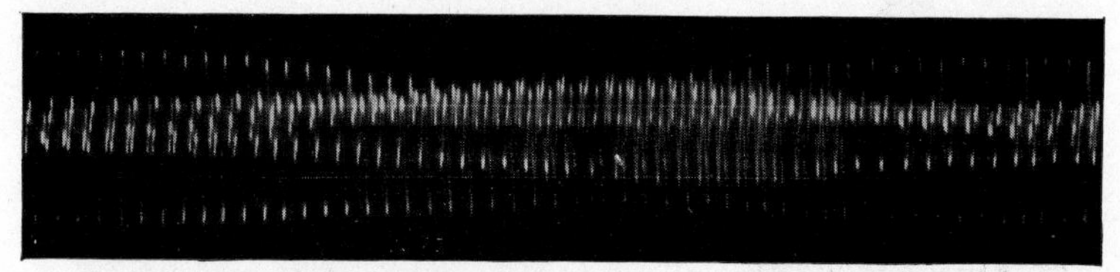

FIG. 11–15. Waveform of violoncello, open A string.

The **double bass** (Fig. 11–17), as its name implies, is the bass of the string section. To it is commonly assigned the important task of supplying the fundamental tones of the harmony. It is the largest of the stringed instruments, and is the sole remaining member of the original viol family. Its shape differs somewhat from

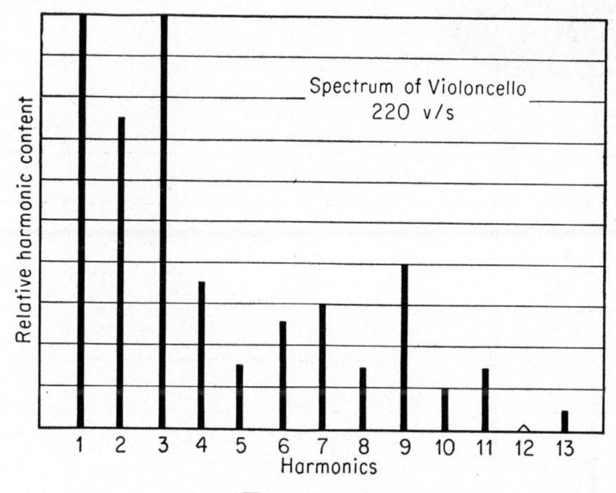

FIG. 11–16

that of the violin and cello, the shoulders being more sloping. Formerly it was made with a flat back, but in recent times the back is commonly curved. Originally this instrument had only three strings, but it now has four, usually E, A, D, G. The fourth string was added in order to adapt the instrument to modern

FIG. 11–17. Double bass. (Carl Fischer Musical Instrument Co., Inc.)

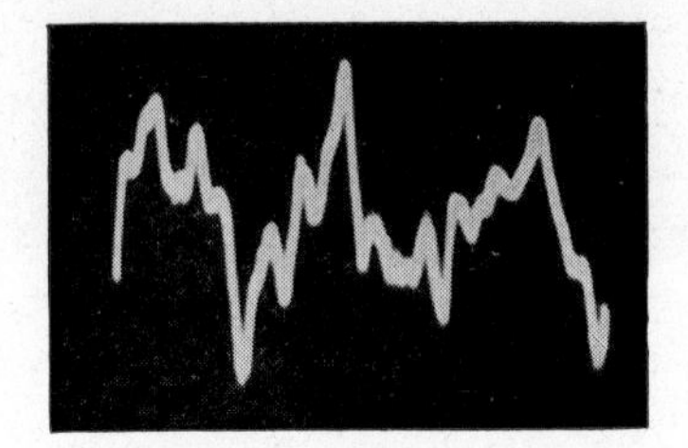

FIG. 11–18. Waveform of double bass, open D string.

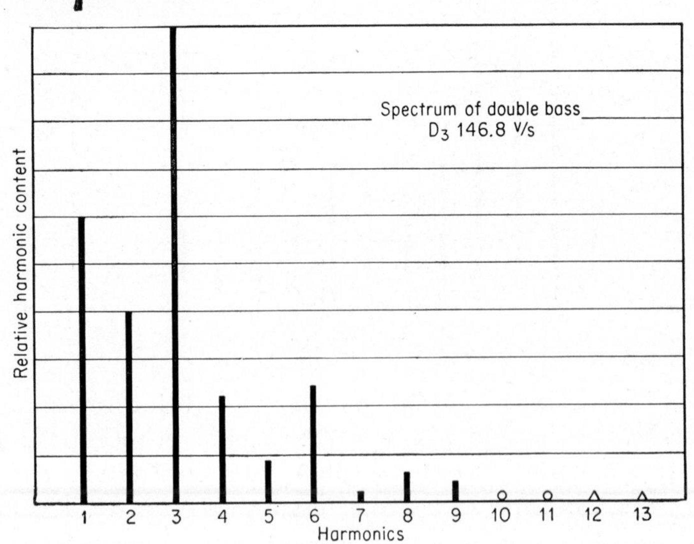

FIG. 11–19

composition. A fifth string is sometimes added and tuned to C in order to supply lower bass notes. The double bass is difficult to play because of the long finger stretches and the relatively great pressure required in fingering. The instrument is tuned in fourths rather than in fifths in order to shorten the finger stretch required.

The waveform of the D string of the double bass is shown in Fig. 11–18, and its sound spectrum in Fig. 11–19. It is to be noted that the lowest note of the double bass is E_1 and the highest B_4.

The double bass supplies the very deep bass parts in orchestral compositions; the modern orchestra commonly employs eight "basses." Occasionally the double bass is given a solo part, as in the introduction to the last act of Verdi's opera *Otello*.

It is possible to produce harmonics on this instrument, though they are infrequently used.

11–3. *The Pianoforte*

The most common of all of the stringed instruments is the one we know as the piano, originally called the pianoforte; it probably has its origin in the ancient harp. As a source of musical sounds it differs from the members of the violin family in several important respects. In the first place, there is a string, or group of strings, for each note, and secondly, the strings are individually **struck** at **fixed points** by felt-covered hammers operated by means of keys. In the case of a given string, the musician can control, within limits, the velocity of the hammer as it makes impact with the string; and, to a certain extent, the length of time during which the excited string may continue to vibrate is also under the control of the player. The strings by and of themselves would produce a sound of very low intensity. Attached, however, as they are to a large and heavy sounding board, whose resonance curve is very broad (Sec. 6–2), they indirectly give rise to musical sounds of relatively great intensity, particularly in the middle and lower register. The sounding board is, or should be, designed to reinforce by resonance those partials which experience has shown should be present in the piano tone. The hammer mechanism is so arranged that, in the middle part of the instrument, it makes impact with the string at a point which is about one-seventh to one-ninth of the length of the string. This arrangement tends to suppress all partials above the seventh.

The strings are of steel; and each is subjected to several hundred pounds tension. From the highest tone down to C_3 three strings of approximately equal length and tension are provided for each note; this is for the purpose of increasing the sound intensity. From C_3 to G_1, inclusive, each note is sounded on two strings, and below this point only one string per note is used. Above C_3 the strings are solid; below this point each string is closely wound with wire. This is for the purpose of attaining relatively great weight per unit length and at the same time preserving some degree of flexibility. The length of any given string depends upon the pitch of the tone to be produced and also upon the type of piano involved. In modern instruments the "harp" varies in length from the extremely small upright models ("spinet" type) to those found in long concert pianos. The advantage of the longer stringed type of instrument is twofold: (1) greater volume is attained, and (2) a more satisfying tone quality is secured—the complement of overtones being more satisfactory in the larger models. However, to attain the same pitch, a given string must be subjected to a greater tension in the case of the larger models. Hence the so-called "harp" (sounding board structure) must be of correspondingly strong construction.

The key mechanism is so arranged that immediately after the hammer strikes the string it flies back out of contact, and a felt damping member is moved into contact as soon as the key is released. By means of the right pedal all of the dampers may be moved away from the strings as long as the musician desires, thus permitting a note or chord to continue to sound while other notes are played. Skillful control of the pedal is an important part of piano playing technique. A second pedal is utilized for the purpose of shifting the entire keyboard mechanism in such a manner that the hammers strike only two of the three wires in each group, thus reducing the volume. On grand pianos a third pedal serves to arrest one damper or more so that a single note or several notes may be sustained. Figures 11–20 and 11–21 show, respectively, the striking mechanism of an upright and a grand piano.

The piano has a wider pitch range than any other instrument

except the pipe organ, the fundamentals extending from A_0 (27.5) to C_8 (4186), slightly over seven octaves.

All tones produced on the piano do not have the same timbre. The lower notes, in addition to the fundamental, have a relatively large number of prominent upper partials, while the tones of the

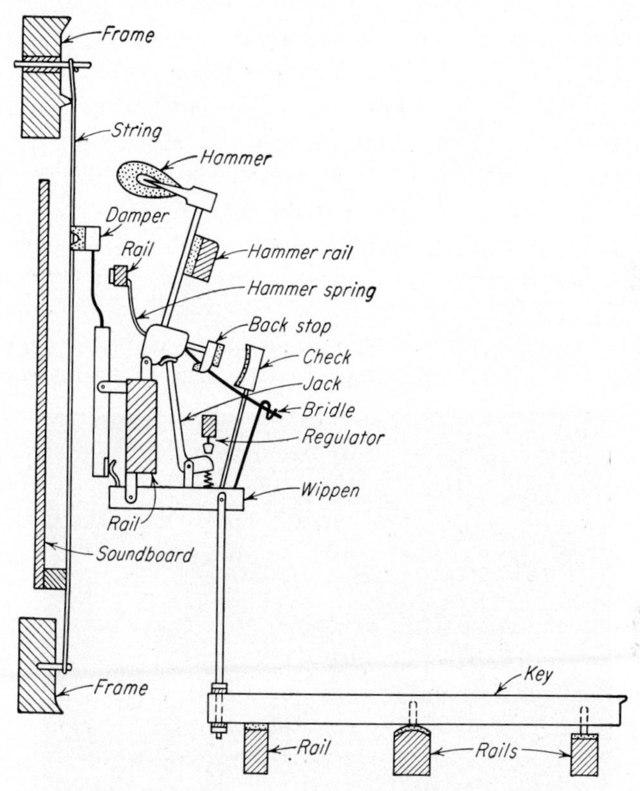

FIG. 11–20. Diagram of striking mechanism of an upright piano. (After H. F. Olson, "Musical Engineering," McGraw-Hill Book Company, Inc.)

middle and higher registers, particularly the latter, are characterized by few and relatively weak overtones. This is evident from the two sound-wave records appearing in Fig. 11–22, and is even more clearly shown by the corresponding sound spectrum charts as set forth in Fig. 11–23.

A careful examination of the records in Fig. 11–22 discloses certain important facts in connection with piano tones. It will be observed that approximately one one-hundredth of a second is required for the sounding board to reach its maximum amplitude of vibration after the impact of the hammer on the string. Almost immediately after the intensity reaches a maximum, the sound

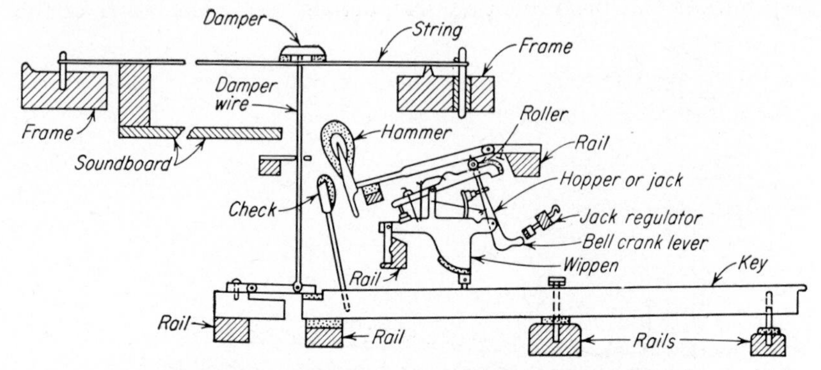

FIG. 11–21. Diagram of striking mechanism of a grand piano. (After H. F. Olson, "Musical Engineering," McGraw-Hill Book Company, Inc.)

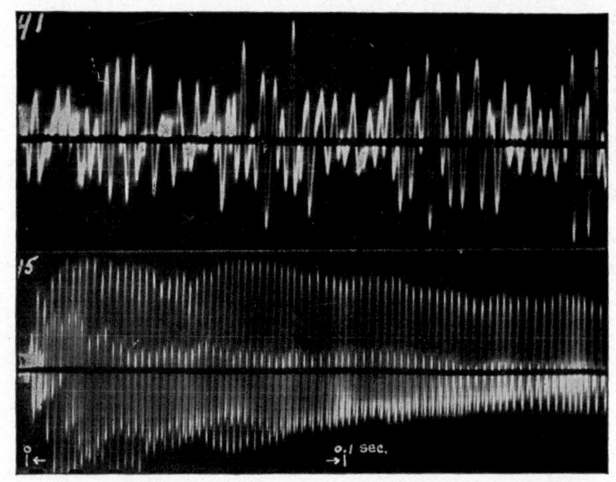

FIG. 11–22. Waveforms of two piano tones. The upper trace is a record of C_1; the lower recording is of C_5.

begins to decay sharply, the rate of decay being most pronounced in the case of the tones of highest frequency. It will also be noted that there is a more or less progressive change in quality as the tone dies out. At first the fundamental predominates, but after a small fraction of a second the fundamental may drop to a lower value than the first upper partial (the octave). The record shows that, in the case of the lower tone, the amplitudes of the upper partials appear to change with respect to one another (sometimes irregularly) as the intensity of the tone as a whole continues to decrease. If the several strings of a given note are not tuned to unison, beats will appear.

From what has been said above it is obvious that the quality of a piano tone is changing continuously as long as it is sounding. In

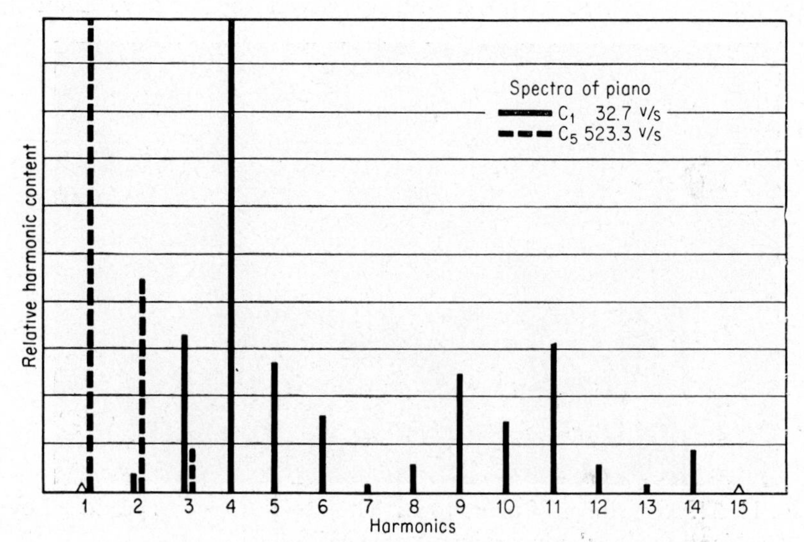

FIG. 11–23. Spectrum of the two piano tones referred to in Fig. 11–22.

fact, experiments conducted by the author indicate that the timbre may undergo measurable change in less than one two-hundredth of a second. This fact probably has a bearing on the much discussed question of "touch."

Many pianists contend that they can produce tones differing in quality by varying the manner in which they strike any given key. It can be demonstrated that this opinion has no basis in fact. Extensive experiments conducted by the writer, and by other investigators, indicate that, if and when the pressure of the stroke is not changed, the musical quality is in no way affected by the manner in which the key is struck. Referring to Fig. 11–24, we see the records of two tones made by a skilled pianist by striking the key in two different ways. It will be seen that the waveforms are identical, so far as general appearance is concerned. Carefully made measurements of these recordings indicate that the two tones are actually identical in quality. What then is the basis of the popular belief about "touch"?

There are probably three aspects of the situation which have a bearing on the case. The first and possibly the most significant factor is the fact that the quality of a piano tone is a function of the magnitude of the initial impulse given to the key. Even a

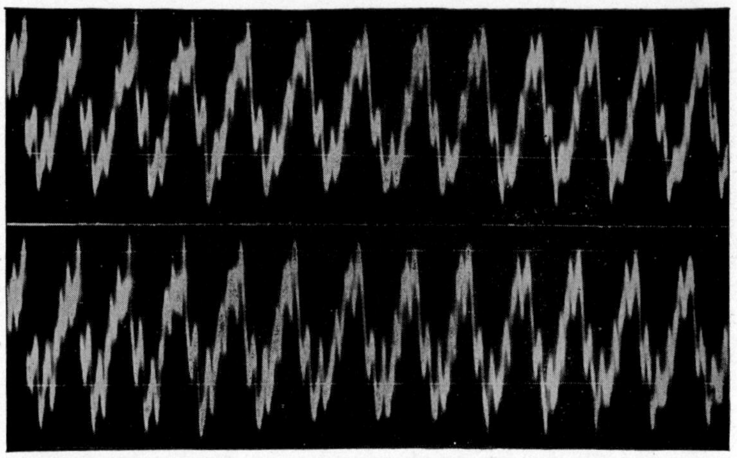

FIG. 11–24. Recordings of piano tones. The upper trace was produced by a "cushion" touch; the lower by a "rebound" touch.

comparatively small change in sound intensity will be accompanied by an appreciable change in quality; the relative amplitude of the upper partials is found to depend upon the loudness of the tone. This is apparent from the examination of Fig. 11–25. Hence changes in "strength of stroke" will be accompanied by some changes in timbre, particularly in the middle register; though only a slight effect is to be noted in the case of the lower and higher notes. It is said that a skilled pianist can evoke as many as twenty different intensities from a string. If then some given note of a chord is struck with a force differing in magnitude from that employed in evoking the remaining notes, it follows that the chord as a whole will exhibit a particular tone color.

Another factor in the case is present when chords are sounded.

Seldom if ever are all the notes constituting a given chord **initiated at the same time;** there will be slight differences in the time at which each note is caused to begin—sometimes consciously introduced. Such differences have a marked effect on the sound of the chord. Above we called attention to the fact that the timbre of a tone changed in an incredibly short interval of time after the strings began to vibrate. It will thus be seen that if, say, the beginning of one note in a chord is delayed slightly, the net total result will be different than it would be if no delay had occurred. Obviously, then, a wide variety of tonal effects are made possible by the way

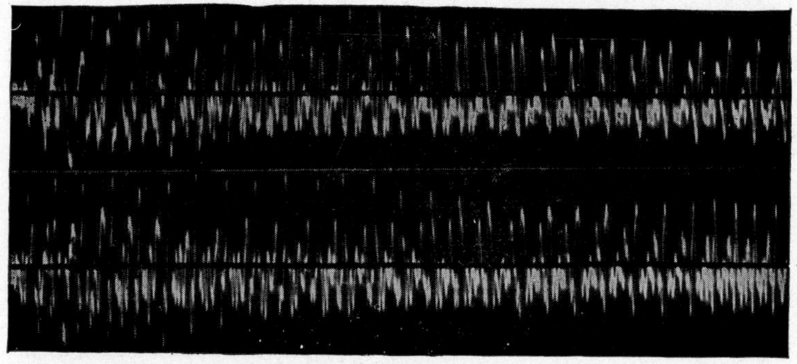

FIG. 11–25. The upper recording shows the waveform of C_3 on a piano when produced by a force of 500 g. The lower trace was recorded when striking the same note with a force of 1000 g.

in which a chord is played. Plainly great skill is required on the part of the performer to accomplish these ends. But, so far as a single note is concerned, no difference in quality can be introduced, except by changing the intensity. We are here dealing with a string of fixed length, excited at a fixed point by a mechanical device. In the case of a single note, given the same magnitude of stroke impulse, a mechanical finger will produce exactly the same quality of tone as would the most skilled musician. Touch probably involves accuracy in judging small time intervals and in correct estimation of slight differences in sound intensity, and the ability to apply such perceptions in the playing of chords.

Any discussion of piano tone quality would be incomplete without a reference to the question of the possible difference in timbre exhibited by the pianos designed and sold by the various manufacturers of such instruments. Is there a difference in pianos comparable to the difference in violins? Our experimental study of this subject shows that different makes of pianos do exhibit definite differences in timbre. This is evident from the spectrum charts shown

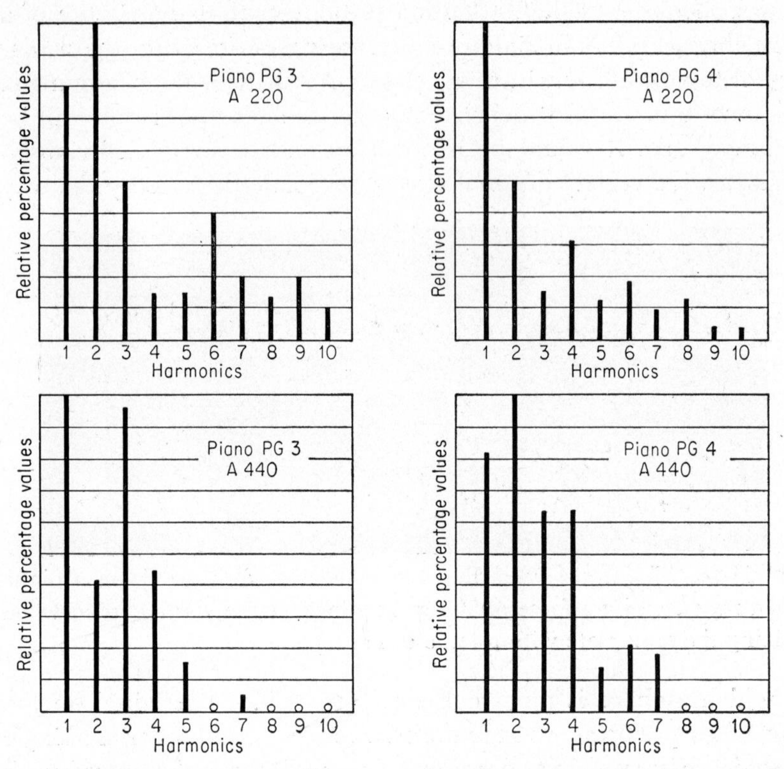

FIG. 11–26. Spectra of two different makes of grand pianos when emitting the tones indicated. Note the marked difference in timbre between the two instruments.

in Fig. 11–26. The two pianos compared were both made by well-known manufacturers, and were of the so-called parlor grand models (7-ft type).

To make the study more complete several additional octaves should have been studied, but the data shown prove that there is an appreciable difference in timbre between various makes of pianos. In Fig. 11–27 there is also shown the difference in tone quality between one of the above-mentioned pianos and a "spinet" (console) model of the same make. While the dynamic range of the small upright model is of course not so great as that of the grand unit, its tonal characteristics compare quite favorably with the large instrument—a result that is somewhat surprising.

In concluding our discussion of piano tone there is one aspect of the situation that should be noted. Actual piano strings are not perfectly flexible components: they exhibit a certain amount of

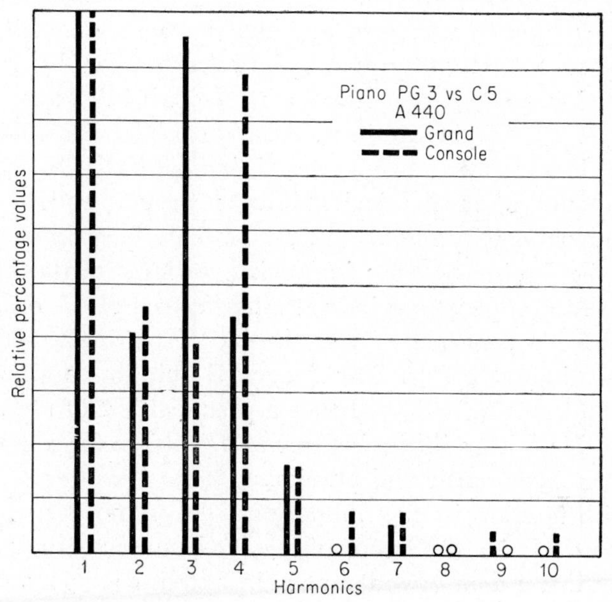

FIG. 11–27. Comparison of the spectrum of a grand and a console (spinet) type of piano of the same make.

stiffness. As a result of this departure from the ideal case the frequencies of the overtones of real strings are not exact multiples of the fundamental mode. This departure from a strict harmonic relationship is referred to as **inharmonicity.** Young and others have studied at some length this question as it applies to pianos. In his latest paper[1] on this subject Dr. Young reports that, in the region of middle C (C_4) of the pianos investigated (six), the inharmonicity was found to be about 1.2 cents (hundredths of a

[1] R. Young, Inharmonicity of Plain Wire Piano Strings, *J. Acoust. Soc. Am.*, vol. 24, no. 3, pp. 269–273, May, 1952.

semitone) for the second mode of the string. He also found that it does not differ greatly for different makes of high-grade instruments. It was noted that below middle C the inharmonicity is small but that above that point it increases rapidly. Dr. Young also observed that for the middle octaves the inharmonicity is appreciably less in the larger pianos than in the smaller ones. It would seem reasonable to assume that the presence of any appreciable amount of inharmonicity would detract from the tone quality of an instrument.

The next question is: Can anything be done to reduce inharmonicity to a negligible quantity? Professor Franklin Miller, of Kenyon College, has suggested[1] a means whereby this desirable end might possibly be attained. Professor Miller has attacked the problem theoretically and his analysis indicates that by applying a small amount of mechanical loading near one end of a piano string, inharmonicity might possibly be materially reduced, if not completely eliminated, thus improving the tone of the individual strings. The computations indicate that a mass of the order of 0.1 g placed a few centimeters from the end of a string would accomplish the desired result. The suggested solution appears to be simple, and it is to be hoped that a practical test of the plan will be made. Here is an intriguing research problem for someone who is versed in both music and physics.

Notwithstanding its few limitations, the piano is a marvelous instrument. Under the hands of a skilled musician the piano becomes an instrument of wide musical possibilities; its scope of tone color, while not as great as that of the violin, is extremely wide, and its dynamic range is unsurpassed.

11–4. *The Harp*

The harp is an instrument of great antiquity. Its modern form is due largely to Sebastian Erard, who in 1810 developed what is known as the double-action instrument. Referring to Fig. 11–28, that portion of the assembly nearest the player includes the soundboard, along the center of which is glued a strip of hard wood carrying the pegs to which the lower ends of the strings are fastened.

[1] F. Miller, A Proposed Loading of Piano Strings for Improvement of Tone, *J. Acoust. Soc. Am.*, vol. 21, no. 4, pp. 318–322, July, 1949.

The wood commonly used in the construction of the harp is sycamore, but the soundboard is usually made of deal. To the under side of the soundboard two horizontal strips are fastened, their function being to cause the soundboard to vibrate in certain modes. The upper ends of the strings are fastened to the curved neck. This member of the assembly also encloses the mechanism by which the pitch of the strings may all be changed a half or a whole tone.

Fig. 11–28. Harp. (Lyon and Healy.)

The pillar is hollow, thus providing space for the rods which serve to operate the pitch-changing components located in the neck. The base of the instrument includes a system of seven levers, in the form of foot pedals, which articulate with the rods passing upward through the pillar. The strings are of catgut, the lowest eight being overwound, somewhat as in the piano. As an aid to the performer all C strings are colored red, and all F strings blue. There are 46 strings, and the instrument as a whole is tuned

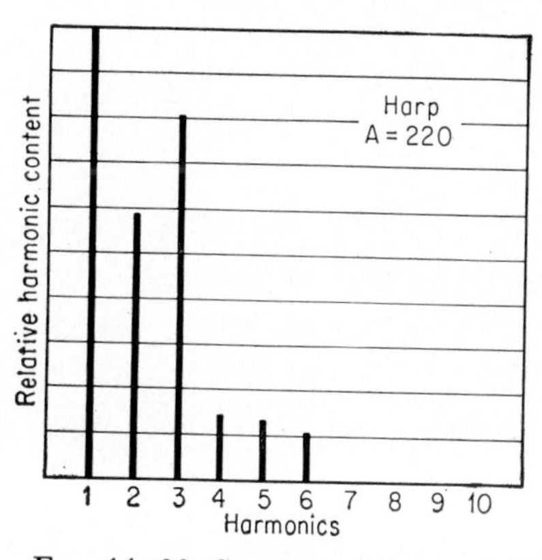

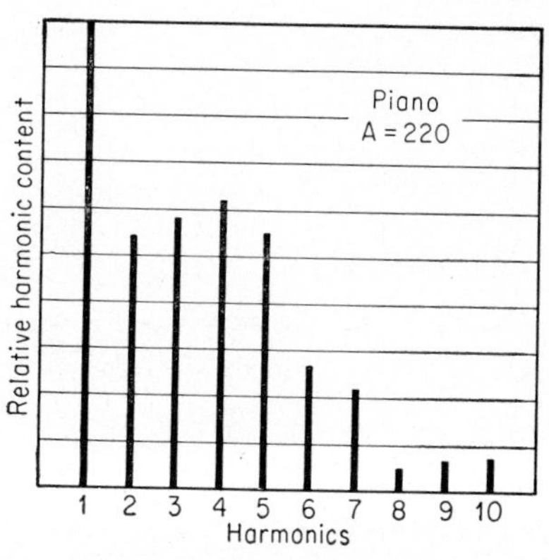

Fig. 11–29. Spectra of harp and piano when sounding the same note.

to the scale of C major. By means of the pedal mechanism, it is possible to play in any key, thus greatly facilitating the rendition of any given score. The Erard double-action harp has a compass of six and one-half octaves; the music is written in both treble and bass clefs. Unlike the pianist, the harpist has relatively wide control of the tone quality of the instrument: the player can vary the manner in which the strings are plucked, and also the point at which the strings are excited. Occasionally the upper partials of the harp are deliberately evoked by a soloist for use in combinations with the flutes and clarinets. The harp is the only plucked string instrument used in an orchestra. Commonly, the modern orchestra employs two harps. Both the physical shape of the instrument and its tonal quality are beautiful. In Fig. 11–29 are seen the sound spectrum of the harp and the corresponding spectrum of a high-grade piano. These spectra should be carefully compared.

12 *Vibrating Air Columns*

12–1. *Vibrations of Air in Closed Pipes*

Thus far in our study of musical sounds, we have considered those sonorous bodies which are **solid.** In all instances the air acted as a medium whereby the energy content of the sound waves was transmitted from the vibrating body to the ear. We are now to examine a case in which a body of confined air is caused to vibrate and thereby become the sonorous body; air will also function as the transmitting medium.

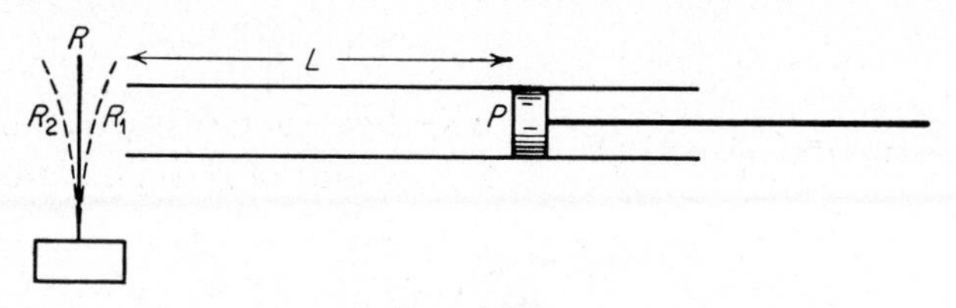

Fig. 12–1

In Sec. 6–2 we saw that a confined body of air might be set into strong sonorous vibration by means of very slight isochronous impulses. It now remains to consider the relations which exist between the frequency of the evoked sound and the dimensions of the sonorous body of air.

Let us assume that we have a column of air enclosed in a tube having a movable piston in one end as sketched in Fig. 12–1. In front of the open end of the tube, we will also imagine, is a reed which is in continuous vibration. Suppose the reed to be moving toward the right, as indicated by R_1. Such a displacement will cause a compression of the air at the mouth of the tube, and such a region of compression will move toward the closed end of the pipe. Upon reaching the piston P, this compression will undergo reflection, and thereafter travel backward toward the open end of the tube. If the tube is the proper length, the compression will reach the open end just as the reed starts to reverse its swing and begins to move toward R_2. The escape of the air from the tube into the region of relatively low pressure will coincide in time with the movement of the reed away from the mouth of the tube. This gives rise to a rarefaction, which in turn will progress down the tube in the direction of P. This, in effect, means that the original compression is reflected as a rarefaction. This rarefaction will in turn be reflected at the closed end as a rarefaction; and as it again reaches the open end, it will undergo reflection as a compression, the reed now being again in motion toward R_1. This process will continue as long as the reed vibrates and L is of the proper length to make the phase of the compressions and rarefactions coincide with that of the reed. Energy is thus transferred periodically from the vibrating reed to the enclosed air. It will be evident from the foregoing outline of the acoustic process that, when such a condition obtains, the length of the tube (L) will be $\lambda/4$; therefore, the

$$\text{Frequency of reed } (f) = \frac{\text{speed of wave disturbance}}{4 \times \text{length of resonant tube}}$$

or

$$f = \frac{S}{4L}$$

The tube is thus resonating to the fundamental of the reed. Now, one may readily find by experiment that there is more than one position of P which will give rise to a resonant condition. For instance, if we make $L = \frac{3}{4}(\lambda)$, resonance is again attained. By proceeding in this manner one could find other adjustments such that the new lengths would equal $\frac{5}{4}(\lambda)$, $\frac{7}{4}(\lambda)$, etc. In each of these cases the resonator has been excited by a generator having a single fixed frequency.

In carrying out any one of the above-indicated experiments, we have caused standing waves in air to be set up, with **displacement** nodes and loops (antinodes) as sketched in Fig. 12–2. In all cases there will be a displacement node at the closed end, motion being stopped by the end wall, and an antinode (loop) at the open end. Bearing in mind that points of zero displacement are points of maximum pressure variation, it will be evident that a **pressure** loop will exist at the closed end. From the foregoing discussion it is evident that the air in a closed tube or pipe may vibrate in several different modes.

Next suppose that our reed generator were capable of vibrating in several ways at one and the same time, and hence of emitting several different frequencies simultaneously. Further, let it be assumed that the tube available is of fixed length and that this length is of such a value that the enclosed air will resonate to the lowest or fundamental sound emitted by the generator. What would be the resonance effects under such circumstances? The tube would respond to the fundamental emission, as in the instance before cited, and as shown diagrammatically in Fig. 12–2*a*. If the reed's emission chanced to have as one of its components a partial whose frequency was three times that of its fundamental, resonance for this sound would also occur and the situation would be as shown in (*b*). If there were present a component whose frequency was five times that of the fundamental frequency, resonance would obtain in this case also, as depicted in (*c*); and so on through the list of partials whose frequencies bear the relation to one another of 1, 3, 5, 7, etc. It may therefore be stated that, in the case of a pipe closed at one end, a series of **odd-numbered** harmonics may be caused to exist. It is not physically possible for the even-numbered partials to exist, because such partials would not be in phase with the exciting agent. In passing it should be noted that the character of the exciting agent is not significant; it may be a reed, a thin ribbon of air rushing across the mouth of the pipe—any means in which, or associated with which, there is a slight vibratory motion having the same frequency as that of the air in the tube.

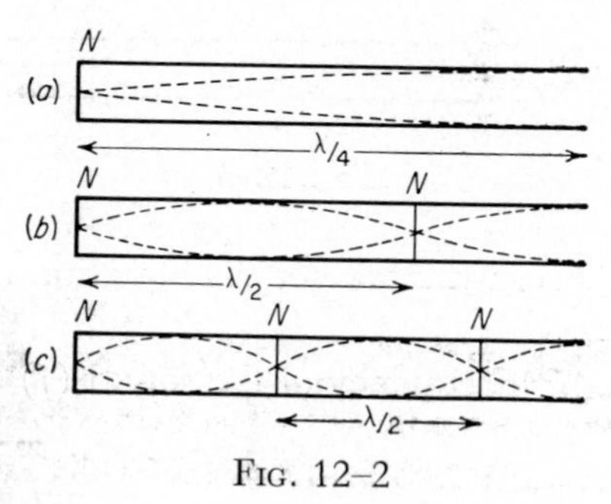

FIG. 12–2

12–2. *Vibrations of Air in Open Pipes*

In the case of pipes which are open at both ends, the situation is somewhat different. From the discussion in the previous section, it is evident that there will be a loop, or antinode, at each end of an open pipe; and the simplest condition under which this could obtain would be as diagrammed in Fig. 12–3*a*. Here the pipe would be emitting its fundamental tone, with a node at the mid-point. In this case the length (L) of the pipe will equal one-half the wavelength of the emitted sound, or

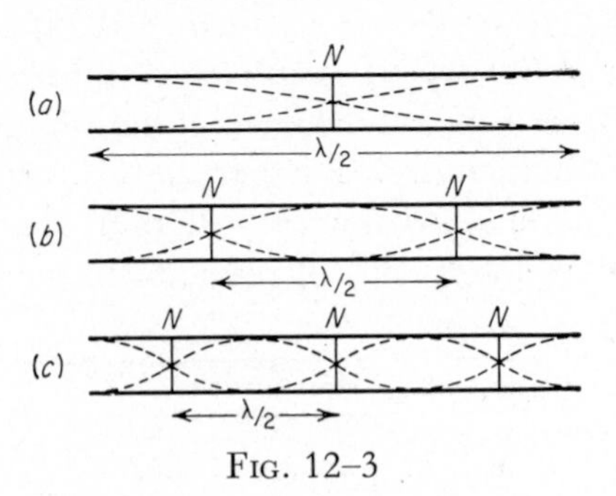

FIG. 12–3

$$L = \frac{\lambda}{2}$$

From this we get the equation that the

$$\text{Frequency of the sound } (f) = \frac{\text{speed of wave disturbance}}{2 \times \text{length of resonant tube}}$$

or

$$f = \frac{s}{2L}$$

Comparing the relations governing the frequency of a closed and an open pipe, it is evident that, for two such pipes of **equal length,** the open pipe will have a pitch **an octave higher** than that shown by the closed unit.

As in the case of the closed type, there are several modes in which the open pipe may vibrate; in each case with **an antinode at each end.** Two such conditions are depicted in (*b*) and (*c*) of Fig. 12–3. In (*b*) we see that the length of the pipe is equal to two half wavelengths while in (*c*) it is equivalent to three half wavelengths. It is therefore apparent that the frequency of (*b*) would be twice that of (*a*); and that of (*c*) would be three times the pitch of (*a*). It follows, then, that, in the case of a pipe open at both ends, **a complete complement of upper partials is possible,** giving the series 1, 2, 3, 4, 5, etc.; this is in contradistinction to the closed pipe situation, where only the odd numbered partials are possible. This is an important distinction, and should be kept clearly in mind as we later examine the practical application of these principles.

Further, it should be carefully noted that, because of the above-indicated difference in the retinue of upper partials, the quality of the tone from a closed pipe will always be different from the timbre of a tone emitted by an open pipe; they cannot possibly be alike in quality. The tone from an open pipe is rich and brilliant, because of the greater number of overtones, while that of the stopped pipe is, relatively, dull.

12–3. *End and Temperature Corrections*

The relation between pitch and length of a sonorous air column, as given in the two preceding sections, is only approximately correct. The length of either an open or closed pipe giving its fundamental note is slightly less than that indicated by theory. In other words, the prime note is lower than that which might be expected from the physical length of the pipe. Certain aerial perturbations occur near the open ends of the pipe which have a slight effect upon the anticipated pitch. The center of a loop does not occur exactly at the end of the pipe but a short distance beyond—a distance which is a function of the diameter of the pipe. For a closed pipe, open at the exciting end, this relation becomes

$$L' = L + 0.58R$$

where L' is the effective length (speaking length) of the pipe in question, L the actual physical length of the pipe being used, and R the radius of the pipe of circular cross section. We may now rewrite our relation for the fundamental frequency of a closed pipe as follows

$$f = \frac{s}{4(L + 0.58R)}$$

The above indicated correction factor is predicated on the assumption that the wavelength is great compared to the diameter of the pipe, which is usually the case. If we are dealing with an open pipe, the end correction must be doubled.

Investigation has shown that the magnitude of the end correction varies slightly with pitch. As one goes up the scale, this correction factor at first rises somewhat and then falls. This means that the higher partials may be sharp with respect to the fundamental and hence may give rise to a certain amount of dissonance. Fortunately if this departure from the strictly harmonic relation (inharmonicity) is at all pronounced, the troublesome partials will be only faintly evoked. Obviously, however, any inharmonicity will, to some extent, have a bearing on the quality of the emitted tone.

Since there is a speed term in the equation given above, and since we have learned that the speed of sound is a function of the temperature of the transmitting medium, it follows that any change in temperature will change the pitch of the sound emitted by a given pipe. It will be recalled that the speed of sound increases as the temperature increases; therefore, the pitch of a sonorous pipe will rise as the temperature rises. In order to compute the natural pitch of any pipe, by means of the foregoing formulas, it accordingly becomes necessary to know the temperature, and to apply the temperature correction, as indicated in Sec. 3–3.

QUESTIONS

1. How long is an open pipe whose fundamental is the same as the second overtone of a closed pipe whose length is 40 cm?

2. What is the length of an open pipe whose fundamental pitch is E_4?

3. An organ pipe has a pitch of 440 at 68°F. What will be its pitch if the temperature of the room changes to 80°F?

4. What would be the frequency of the third harmonic of a closed pipe whose length is 8 ft, neglecting end corrections? Of the sixth harmonic?

5. Why must the end correction be doubled in the case of an open pipe?

13 *Wind Instruments*

13–1. *Aeolian Tones*

Before taking up a detailed study of wind-blown instruments it will be useful to give consideration to certain basic acoustic phenomena that have a bearing on the generation of tones in such instruments.

When there is relative motion between a solid body of small dimensions and a fluid, eddies or vortices are formed behind the moving body. The existence of such eddies, when a liquid is the medium, is shown in the illustration appearing as Fig. 13–1. It will be noticed that these vortices alternate on either side of the path of the solid object. If a stick, for instance, is being drawn through still water it will be found that a small shock is given

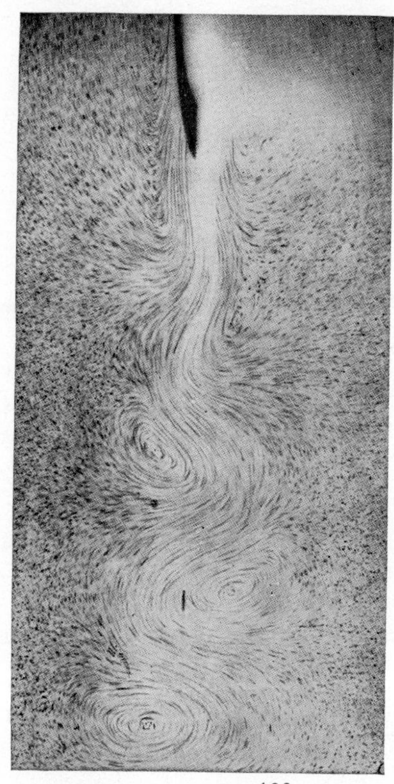

Fig. 13–1. Eddies formed when a body is moved through a still liquid. (From a paper by G. J. Richards in *Phil. Trans. Roy. Soc.*, October, 1934. Reproduced by permission.)

to the stick as each eddy is formed. The neighboring portion of the liquid also receives a slight pulse.

Quite similar phenomena occur when there is relative motion between air and a solid body. In the case of air the formation of each eddy likewise imparts a small impulse to a wire or other object past which the wind may be blowing. If the rate of formation of the eddies happens to correspond to the natural period of a wire, for example, it will be found that the wire will be thrown into vigorous vibration, with the result that an audible sound is heard—the wire "hums" or "sings," we say. Sounds thus generated are known as **aeolian tones.** Such sounds are heard when the wind blows through tall grass or the trees. One of the ancient musical instruments known as the aeolian harp operated on this principle. Strangely enough the pitch of an aeolian tone does not depend upon the tension of the wire or its length, but is instead a function of the relative speed of the air and the object—as the speed of the air increases the pitch rises.

By this time the reader is perhaps wondering what all this has to do with the generation of musical sounds in any modern musical instrument. The answer is that there is an intimate connection between the phenomena above mentioned and the production of a tone by a flute or an organ pipe. We shall now consider that aspect of the situation.

13–2. *Generation of Musical Sounds by the Vibration of Air Columns*

In Chap. 12 we discussed the laws that govern the manner in which the air in closed and open pipes vibrates. We now address ourselves to the question of how the vibration of such air columns is initiated and maintained. There are two methods by which an air column can be caused to vibrate: (1) by reeds and (2) by the movement of air against an edge. Later we shall consider the first method, but for the moment we are concerned with the latter method mentioned.

If a stream of air issuing from a narrow slit is caused to impinge on a sharp edge, the physical conditions will be essentially the same as when the wind blows against a telephone wire. Small aerial vortices will be formed on both sides of the wedge and these eddies will give rise to a faint sound known as an **edge tone.** As in the case of the aeolian tones, the pitch of the edge tone will depend upon the speed of the air stream. Now if there is associated with this mechanical edge a column of air, this enclosed air will be thrown into vibration as a result of the formation of the aerial eddies. If the frequency of the eddy formation chances to correspond approximately to the natural frequency of the air column, resonance will occur, and a relatively loud sound will result. The eddies plus the air column constitute an acoustically coupled system, and once the air column is in vigorous vibration, it will, to some extent at least, tend to cause the eddy frequency to coincide with its own period.

A flue organ pipe constitutes an assembly corresponding to the arrangement above suggested. A sketch of such a unit appears in Fig. 13–2. Air at a pressure of the order of .2 lb/sq in.* is caused to issue from a wind chest as a thin aerial ribbon. As it strikes the lip this stream of air breaks up into a series of eddies. The formation of these eddies within the pipe initiates pressure pulses having a fairly definite frequency. The fact that such eddies exist physically is clearly shown by the striking pictures originally produced by Carrière,[1] and reproduced in Fig. 13–3. In examining this illustration it is to be noted that the ribbon of air also swings back and forth across the edge of the wedge (the "lip"). There is some difference of opinion among those who have studied this phenomenon as to the process by which the air column is actually excited. However, Carrière found that when no vortices were formed no sound is heard. It is possible that the swinging of the air ribbon, due to the formation of the eddies, is also a factor in causing the pipe to "speak." In any event, the aerial perturbations occurring

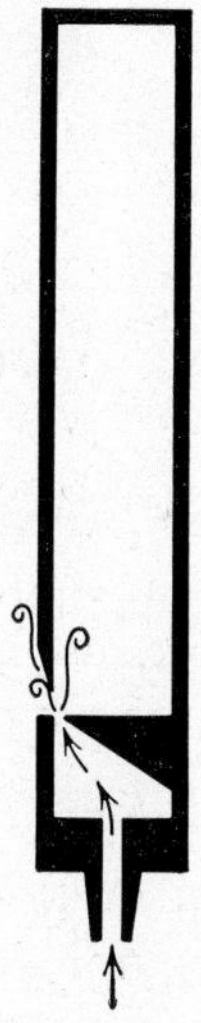

Fig. 13–2. Diagram of a flue organ pipe.

* In practice, organ air pressure is often expressed in terms of inches of water as measured by an open-tube manometer.

[1] Z. Carrière, *J. phys.*, vol. 6, no. 52, 1925, and vol. 7, no. 7, 1926.

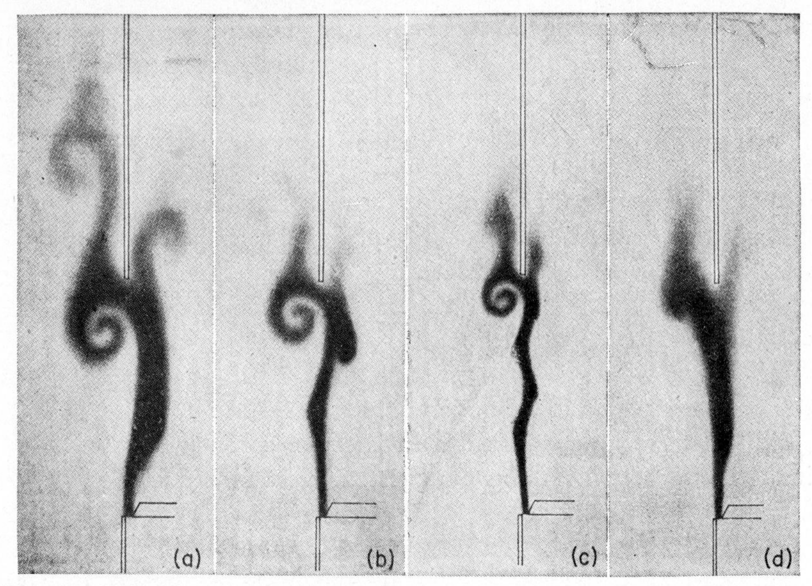

Fig. 13–3. Vortices formed at the mouth of a flue organ pipe. The wind pressure in (*a*) is normal, but is successively lower in the other three cases. (From a paper by Z. Carrière in *J. phys.*, vol. 6, p. 52, 1925. Reproduced by permission.)

at or near the lip serve to initiate and to maintain the pulses of compression and rarefaction required to cause the air in the tube to vibrate and thus emit a sound, according to the laws outlined in the last chapter.

13–3. *Organ Pipes*

The modern organ is a collection of a number of instruments, and each instrument, or "stop," is composed of a group of air-blown pipes. Daniel Bernoulli is commonly credited with having been the first to discover the laws which govern the notes emitted by the pipes used in an organ. Later, Mersenne, Savart, Helmholtz, Rayleigh, and Koenig, by theoretical and empirical investigations, extended our knowledge of these laws and their practical limitations. Some of these laws we have already examined in the preceding chapter. It may be added here that when the cross-sectional size of a pipe becomes comparable with its length, the formulas connecting pitch and length break down. Mersenne was able to lower the pitch of a pipe seven whole tones by holding the length constant and increasing the diameter from one-fourth in. to 4 in. The work done by Mersenne was extended by Savart to include pipes of various forms. As a result, we have what might be designated as the law of Mersenne and Savart, which is to the effect that: "Two similar pipes having similar embouchures emit notes whose pitch is inversely proportional to their lineal dimensions." This means, for instance, that the fundamental tone of a 12-in. pipe of square cross section (say 4 in. on a side) will give rise to a note that is an octave lower than a similar 6-in. pipe which is 2 in. on a side. Organ builders have worked out several empirical formulas for use in the design and construction of organ pipes. M. Cavaillé-Coll, a celebrated French organ builder, makes use of the following relations:

For cylindrical pipes of diameter d

$$L' = L + (5/3)d$$

For rectangular pipes of depth a

$$L' = L + 2a$$

where L' and L have the same significance as before, and a is the length of one cross-section side.

Organ pipes are of two general types: (1) flute (sometimes called flue) pipes, and (2) reed pipes. The flute organ pipes follow, in a general way, the design sketched in Fig. 13–2, though there are various modifications of this structural plan. For instance, such pipes are made in various shapes, as shown in Fig. 13–4, and certain auxiliary components are sometimes attached near to, or even in, the mouth; in certain cases the edge of the lip is notched. By means of such structural details the harmonic content of a given pipe is modified in such a way as to yield a tone having more or less definite timbre characteristics. Such terms as bourdon, gamba, and viole céleste, etc., are used to designate the particular tone qualities thus secured.

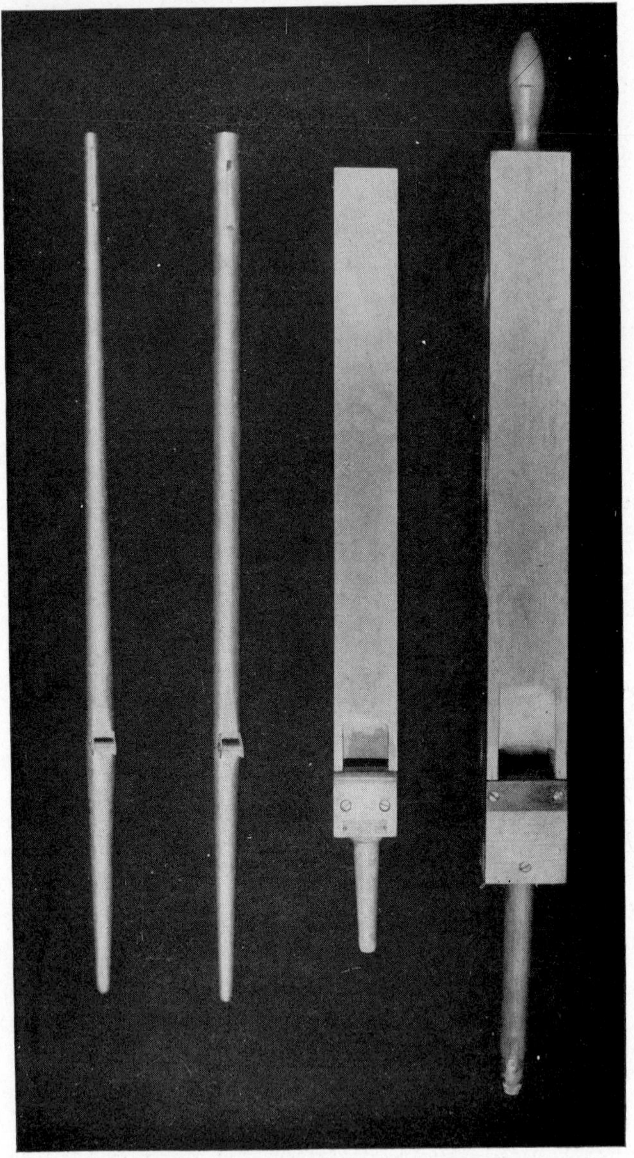

Fig. 13–4. Representative flue-type organ pipes. *Left to right:* viole céleste, gamba, flute, bourdon.

In general it may be said that the tonal quality of the sound from an organ pipe depends on such factors as:

1. Whether the pipe is open or closed
2. The scale of the pipe
3. The shape of the pipe
4. The magnitude of the wind pressure
5. Method of excitation (by edge or reed)
6. Thickness and nature of wall material
7. Shape and size of the mouth and associated parts

Referring to the second item in the foregoing list, the term "scale" designates the ratio between the diameter and the length of the pipe. For instance the gamba pipe shown in Fig. 13–4 would have a scale of something like $1/22$ or 0.045, while the scale of the bourdon would be about 0.16. The scale of a pipe has a marked effect on the tone quality. The larger the scale value the more difficult it is to excite the higher partials, hence a pipe of large scale has relatively few upper partials. By the same token a slender pipe (small scale) will yield a tone rich in overtones. Further, since the end correction varies to some extent with the frequency (Sec. 12–3) it will be evident that this factor also has a bearing on the quality. In fact, the scale and end correction factors are interrelated in their influence on quality. One is tempted to discuss in detail the characteristics of the various organ pipes, and how the timbre is attained in each case, but space forbids. Those interested in following up this and related questions should consult a work by Dr. William H. Barnes entitled, "The Contemporary American Organ," 5th edition.

The majority of pipes are made of zinc, tin, or some alloy. Some pipes, particularly the larger ones, are made of wood.

The tuning of open flute pipes is adjusted by means of a narrow strip cut in the pipe wall near the top. The cut-out portion is rolled up or unrolled, thereby lengthening or shortening the "speaking" length of the tube. In the case of stopped flute pipes tuning is effected by means of a tight-fitting adjustable plug inserted in the upper end of the pipe.

Thus far we have been confining our attention chiefly to the flute type of pipe. The **reed pipes** of an organ differ in certain important respects from the type of pipe which we have just been considering. Instead of a thin ribbon of air functioning as a sound generator, we have a metal reed serving as the means of initiating condensations and rarefactions in the associated resonator. Two general types of reeds are possible: (1) free reeds; and (2) striking, or beating, reeds. In the former, the vibratile part of the reed is slightly smaller than the aperture over which it vibrates. The reed organ, the harmonica, and the accordion are examples of free-reed instruments. In the second type the reed is slightly larger than the associated aperture. The clarinet makes use of a reed of this type. Originally both free and striking reeds were incorporated in organs, but in recent years the tendency is to use the beating type of generator exclusively. In Fig. 13–5 the essential components of a reed organ pipe are shown diagrammatically. *S* is a metal tube having a part of one of the walls cut away. *R* is the reed, usually of brass, held in place by means of wedge *B*. A wire *W*, whose lower end rests against the reed, provides a means whereby the reed is tuned. The reed is curved slightly outward from its base throughout its length; it therefore does not rest in contact with the metal tube, called the *shallot*. Air entering the chamber *C*, as shown, rushes through the narrow opening between the reed and the shallot, thus causing the reed to vibrate. The result is that the opening is periodically closed and opened; puffs of air enter the pipe *P*. Each puff of air causes a condensation to proceed upward in the tube. This is reflected, at the open end of the pipe, as a rarefaction. By the time this rarefaction reaches the reed the reed opening has been closed, and the reed will momentarily remain in contact with the shallot due to the presence of the rarefaction. The rarefaction will accordingly be reflected a second time, but in this case without change of phase. Upon arrival at the open end, the phase will be reversed, and the rarefaction will be reflected

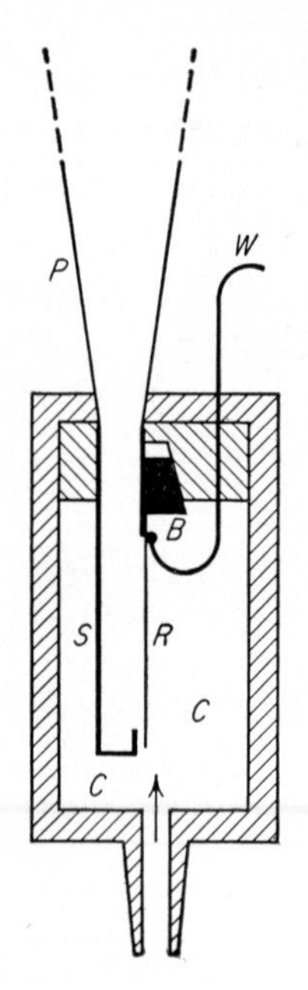

Fig. 13–5. Diagram showing construction of reed organ pipe.

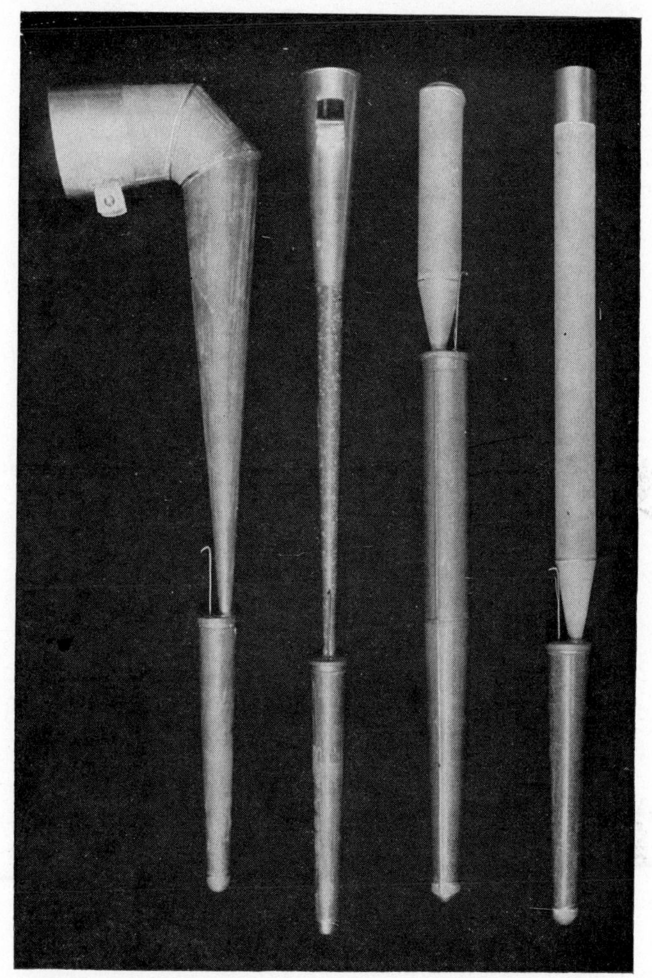

FIG. 13–6. Typical reed organ pipes. *Left to right:* tuba, oboe, vox humana, clarinet. The curved shape of the tuba pipe is not significant.

as a condensation. Upon arrival at the reed end, the pressure due to the condensation plus the force of restitution of the reed itself will cause the reed to move away from the shallot, and thus a cycle of operations is completed. This process will be repeated as long as air is supplied to the chamber C. Two important facts are apparent from the above discussion: (1) The pulse travels four times the length of the tube during the time required for the reed to execute one complete

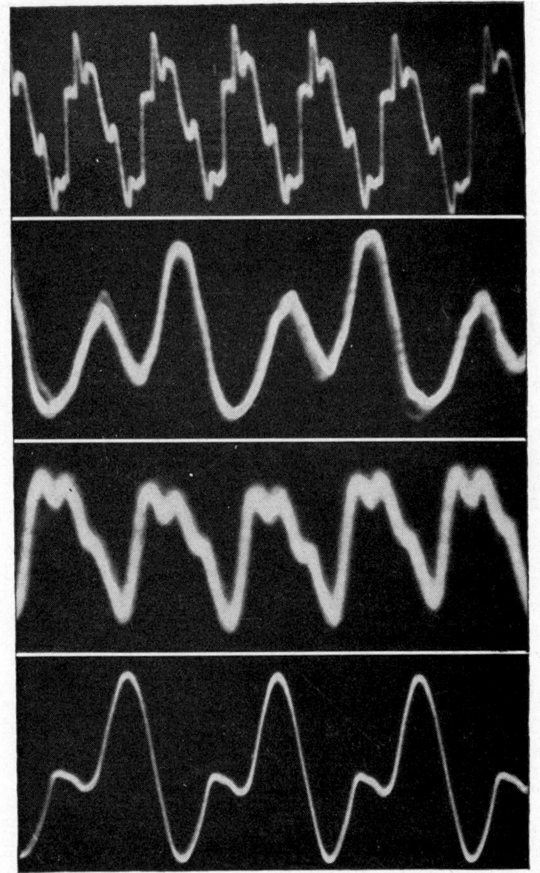

Fig. 13–7. Waveforms of organ pipes. *Reading down:* clarinet, A 220; viole, A 220; harmonic flute, A 440; diapason, A 220. Heaviness of trace not significant.

vibration; hence the wavelength is four times the length of the pipe; (2) The air in the resonating pipe is mechanically closely coupled with the reed. This latter fact means that the "speaking" frequency of a reed pipe is determined by the joint effect of both the generator (the reed) and the resonator. It has been determined by careful experimentation that, generally, the speaking frequency of a reed pipe is lower than that of either the reed itself or its associated pipe. If the reed is too stiff, the generator may impose its natural frequency on the pipe, but in the case of organ pipes the reed is of thin metal, thus bringing about a condition of reciprocal influence. In practice the reed and pipe are arranged to have nearly the same frequency; this condition makes for a more satisfactory tone, and a "quick speaking" pipe assembly.

From (1) above it would appear that a reed pipe might be considered to be a closed pipe, with a node at the reed end. Actually

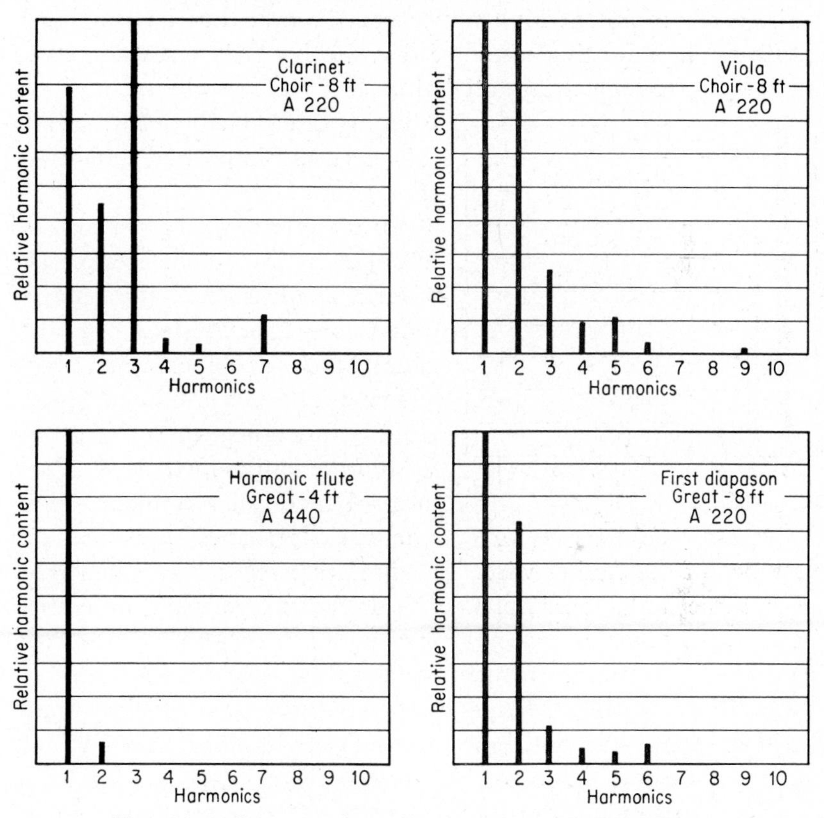

FIG. 13–8. Spectra of organ tones shown in Fig. 13–7.

such a pipe does not function as a closed pipe, as will be evident from the sound spectrum (Fig. 13–8) of a clarinet pipe. In practice both orchestral and chorus reeds (discussed below) show even as well as odd overtones.

A reed pipe is tuned by altering the natural period of the reed, rather than by altering the length of the pipe. The lower end of wire *W* (Fig. 13–5) presses against the reed, and when moved up or down serves to lengthen or shorten the vibratile part of the reed, and thus changes its period.

By making the vibratile member assume a particular shape, the reed is prevented from closing abruptly; it closes with something of a rolling motion, closing first at the base end. If the moving air were to be abruptly stopped, the resulting tone would contain a retinue of prominent high upper partials and thus have a harsh quality. The adjusting of the reed and other adjacent elements of a pipe to produce the desired timbre is called **voicing,** and the shaping of the reed constitutes one of the most important phases of that process.

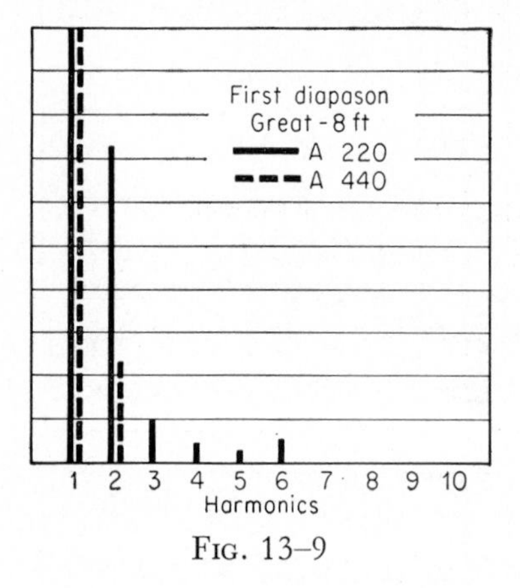

Fig. 13–9

In the organ, reed pipes may be classified in two main groups: (1) **chorus reeds,** such as the trumpet and tuba; (2) **orchestral,** or **imitative, reeds,** as for example oboe, clarinet, and vox humana. Figure 13–6 is a photograph of a group of representative reed pipes. They are all made of soft metal. Note the various shapes, and also the presence of the tuning wire. The imitative reeds are designed to bear some tonal resemblance to the corresponding orchestral instruments, but the sound spectrum of these stops bears little resemblance to the instruments after which they are named.

One is tempted to discuss in detail the characteristics of the various organ pipes and how the timbre is attained in each case, but space forbids. However, we cannot close our all-too-brief discussion of organ pipes without at least referring to that group of stops commonly spoken of as the "strings," of which the viole d'orchestre is a conspicuous example. The so-called string tones are secured from flute pipes by modifying the physical relation of the elements which determine the shape and size of the embouchure, by giving special contours to the lip, and by other special structural modifications. In Figs. 13–7 and 13–8 will be seen a record of the waveform of representative flue and reed pipes, together with the corresponding spectra. These waveforms and spectra should be carefully compared, and the essential differences in the harmonic content noted.

In this connection it should also be noted that the timbre of organ tones on a given stop may differ radically depending on the position of the note in the musical scale. For example the 440-A on a diapason stop has a decidedly different tonal quality from the 220-A in the same rank of pipes. This difference is strikingly shown by the two corresponding spectra shown in Fig. 13–9. It is obviously important that both the composer and player recognize this significant fact.

13–4. *The Organ*

In considering the organ as an instrument it may be said that there are four general groups of pipes: (1) the diapasons, (2) the flutes, (3) the strings, and (4) the reeds. The diapason tone serves as the basis of organ tone structure. It is this group of pipes that distinguishes the organ from all the other musical instruments. Each of these main groups is made up of a number of subgroups, each having the same general tone characteristics but differing in certain details. Thus an organ is an assembly of a number of instruments, each such instrument (a suborgan) being composed of groups of pipes called ranks, or stops. The number of stops in an organ varies from something like ten to more than a hundred. The various stops may be connected to one of the keyboards by means of draw knobs, located on either side of the manuals, or by the use of levers located above the upper keyboard. The operating mechanism of an organ is so constructed that the organist can make use of any particular individual stop ("voice") or a group of stops. The stops in turn are assembled in a number of major groups, each group constituting essentially a separate organ, the number of such units ranging from two to eleven. Provision is made, however, so that the several organs may be played simultaneously. This is accomplished by what are known as "coupling" mechanisms. Each single suborgan is played from its own keyboard. One of them, the pedal organ, is played by the feet and the others by the hands, the keyboards of the latter being known as manuals. The several organs are called Choir, Great, Swell, etc.

In dealing with the several stops one often hears such terms as "8-ft" pitch. By that is meant that the longest pipe (open type) of that stop is approximately 8 ft in length, and hence its fundamental tone is middle C (C_4). Many organ stops are of 8-ft pitch. A 4-ft stop would be an octave higher and a 16-ft stop an octave lower than the 8-ft group.

Certain of the organ pipes yield sounds of relatively low intensity, while others are capable of producing tones whose intensity is high. But the over-all control of the sound level is brought about by means of adjustable shutters that form one side of the enclosure

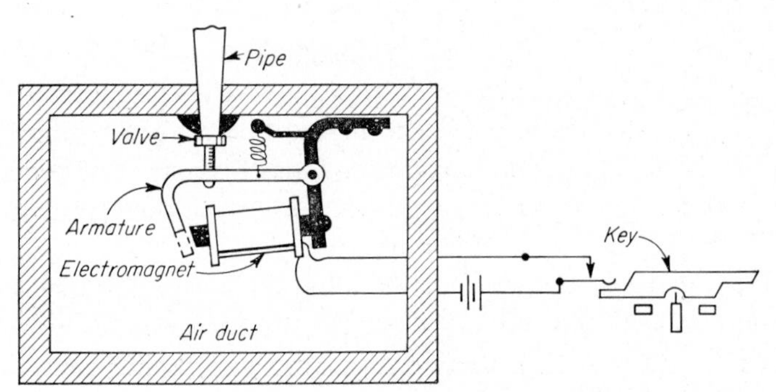

FIG. 13–10. Diagram of organ valve mechanism.

in which one or more of the organs are housed. By means of these shutters, which are controlled by a foot pedal, the volume of sound can be smoothly modulated. This organ is the one designated as the swell unit.

The air supply for the organ as a whole is provided by means of a large high-speed fan driven by a suitable electric motor. The fan, or "blower," as it is usually called, forces air into a wind chest from which air ducts lead to the several groups of pipes.

The *modus operandi* by which air from the wind chest is admitted to the individual pipes, when the keys are depressed, is shown diagrammatically in Fig. 13–10. There are two general schemes for accomplishing this end, one known as the *electropneumatic* action and the other as the *full-electric* plan. The electropneumatic system is the one most commonly used and the sketch, somewhat simplified, indicates the essential elements involved in that plan. A study of the diagram will disclose that the depression of the playing key operates an electrical relay which in turn activates a pneumatic relay that serves to admit air to the attached pipes. In the case of full electric operation the pneumatic relay is replaced by an electric relay unit.

It is also of considerable interest to observe that the tone quality of different organs differs widely. The fact is quite apparent when one examines the spectrum charts appearing in Fig. 13–11. In this connection it should be pointed out that two like pipes, two bourdons for instance, made by the same manufacturer and voiced by the same expert, may, under different ambient acoustical conditions, yield quite different sound spectra. In many churches the acoustical surroundings seriously detract from the optimal tone quality of one or more ranks of pipes. It is therefore of the greatest importance in designing a new church building that the location of the organ pipes be given careful consideration. Unless the architect and the organ builder collaborate closely in such an undertaking the final results may prove to be disappointing.

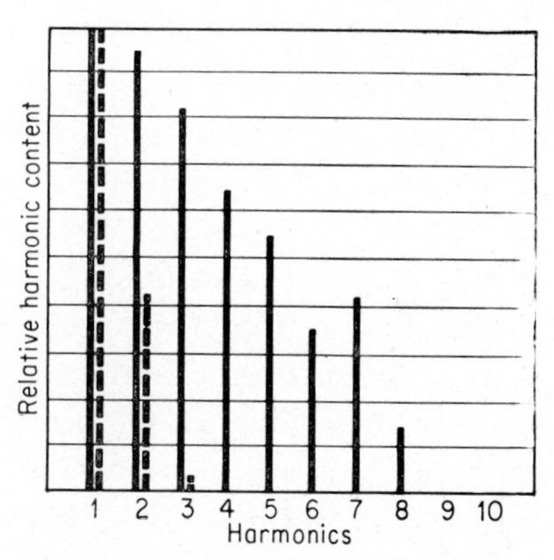

FIG. 13–11. Spectra of a given stop on two different organs. Note the marked difference in timbre.

In this general connection there is a related factor that is of considerable importance. We refer to the matter of the acoustical treatment of a church interior, particularly the nave. Now that more attention is being given by architects to the matter of the reduction of excessive reverberation, organists are showing considerable apprehension concerning the possibility of overdoing the application of sound-absorbing material to the walls and ceiling of church interiors, and its possible effect on organ tone. At a recent conference of organists this question was given considerable attention.

In his book "The Contemporary American Organ," before referred to, Dr. Barnes in the last chapter, under the caption Requirements for Good Church Acoustics, takes what would appear to be an extreme position in this matter. The author has had some experience as a consultant in connection with the acoustics of church interiors and is personally acquainted with several of the leading acousticians in the country. To the best of our knowledge and experience every reputable acoustical engineer gives careful and sympathetic consideration to the musical aspects of the situation when dealing with church interiors. Dr. Barnes, apparently, feels otherwise. The acoustician faces a diffcult task in dealing with church interiors; he is called upon to arrange conditions so that the pastor can be heard and understood and also so that the organist and the singers may be given full opportunity for correct expression. There may be cases in which a compromise may be necessary, but the organist is never deliberately penalized. Further, it may not be out of place to point out that the tone quality and the expression that the organist hears when sitting at the console is not necessarily the same as that experienced by the members of the congregation, as an objective test will readily show. Sympathetic cooperation between the organists and the acousticians is necessary in order that the best over-all results may be attained.

In concluding our discussion of the organ it might be said that the unique thing (musically) about an organ is that it is possible to play, simultaneously, the various ranks of pipes, each having its own individual tone characteristics. An organ may thus combine what amounts to many instruments and thus produce a composite tone color of great beauty and wide dynamic range. In this respect the organ differs from all other musical instruments. In the case of an orchestra some seventy-odd players are involved in securing the combined effect. At the console of an organ a single performer attains a corresponding result. Notwithstanding certain mechanical and tonal limitations, the modern organ is a magnificent instrument. Organ music is particularly adapted to serve as a medium whereby man may express those aspirations and emotions which are inseparably connected with his religious life.

13–5. *The Flute*

Its great age, its essential mechanical simplicity, and the tender beauty of its tone color combine to make the flute the most interesting of the orchestral wind instruments. It consists of a metal or wood tube about 26 in. in length with a ¾-in. bore, open at one end (Fig. 13–12). The embouchure is an oval opening near one end. The flute shown in the illustration is provided with a series of 16 openings in the tube wall, 11 of which may be closed directly by seven fingers and one by the left thumb. The four additional openings may be opened or closed by means of suitably arranged keys. The openings serve to modify the effective length of the resonant air column.

So far as the generation of a single musical tone is concerned, the flute functions in the same manner as a flute organ pipe. The player causes a stream of air to impinge on the sharp edge of the embouchure, thus giving rise to an edge-tone. The pitch of the

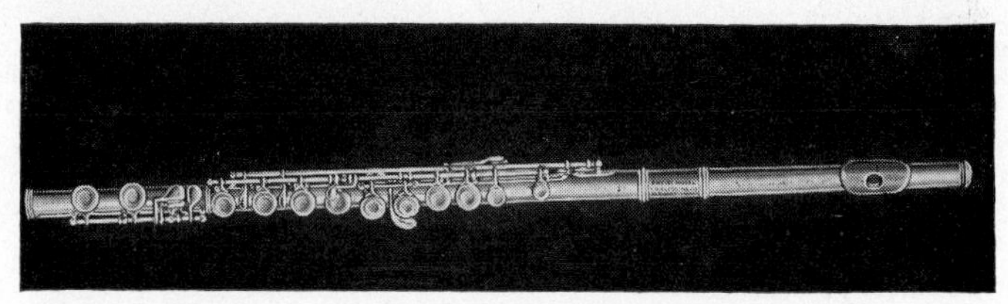

FIG. 13–12. Flute. (Carl Fischer Musical Instrument Co.)

edge-tone is governed by the velocity of the air from the player's lips. As in the flute organ pipe, the enclosed air column is thrown into vibration by means of the aerial reed, and resonates at the frequency determined by its operating length. By control of the lateral openings, and by changes in the air pressure, the flute can be made to cover a range of three octaves beginning with C_4 and extending to C_7.

The tones of the flute are characterized by an extremely low harmonic content. Figure 13–13 shows representative waveforms of a high and a low note, and Fig. 13–14 depicts the corresponding spectra of these two tones. It will be observed that the sound in the upper register contains only a small percentage of a single upper partial. When sounded softly the notes in the upper register frequently have no overtones present. In the middle register there are traces of upper partials, but the fundamental predominates. In the lower register the tone is characterized by a somewhat greater harmonic content. If the lower register is sounded loudly, the tone color changes decidedly, the octave and the fourth harmonic predominating.

How is it that an open-pipe type of instrument yields a tone having so low a harmonic content? No completely satisfactory

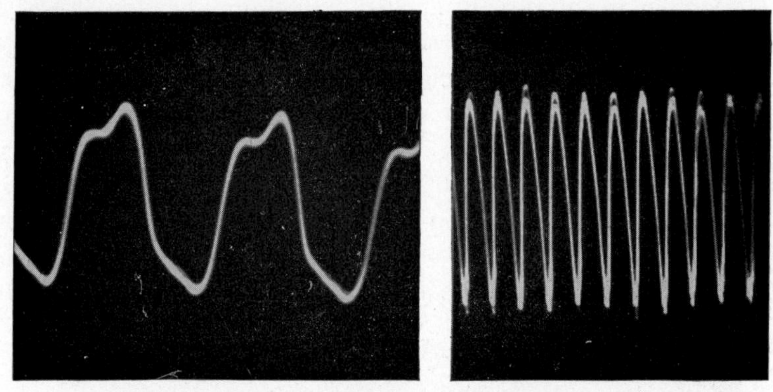

FIG. 13–13. Waveforms of two flute tones.

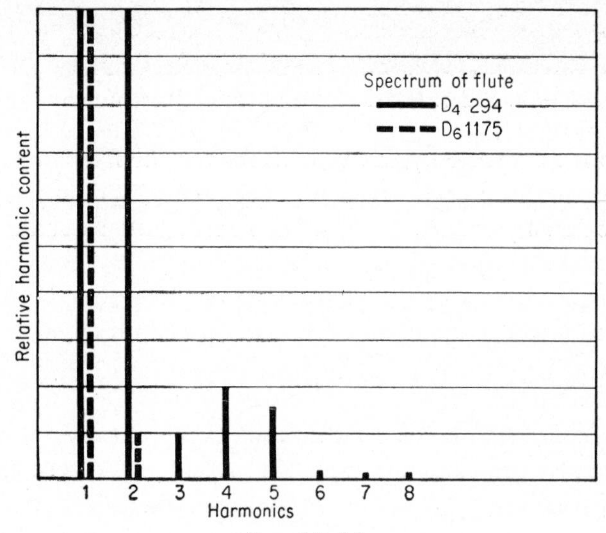

FIG. 13–14

answer to this question is possible in the present state of the art of musical acoustics. However, certain aspects of the case may be examined with profit.

We have seen in the case of strings that, for the higher tones, there are few upper partials. Theory shows that, not only in the case of strings, but for high frequencies in general, the higher harmonics have only a small amplitude; and this is particularly noticeable in the case of sonorous air bodies. In the case of the flute the nature of the tone-generating process probably tends to make the higher upper partials inharmonic and hence only feebly emitted. The nature of the material of which the tube is made also appears to have an appreciable effect upon the harmonic content of flute tones. Miller, who has exhaustively examined the functioning of the flute, concludes that instruments made of silver and gold have a greater harmonic content than those made of wood. It would appear that the density of the material and its elasticity are both factors in the case. There are those who feel that a metal instrument yields a "liquid" tone, particularly suitable for solo work; while one made of wood (cocus, ebony) exhibits a rich mellow tone which is useful in orchestral work. The weakness of the upper partials makes the instrument particularly useful in connection with accompaniments for the soprano voice, as, for instance, in the flute obbligato in the Mad Scene from Donizetti's *Lucia*, and in "Lo, Here the Gentle Lark" by Bishop. Indeed one of the most surprising features about the tone color of a flute is its remarkable birdlike timbre, particularly in the highest register where the tone consists almost entirely of fundamental. It is frequently used in solo parts, a beautiful example being the composition "Wind Among the Trees" by Briccialdi. Three flutes are commonly employed in the modern symphony orchestra.

13–6. *The Piccolo*

The design and construction of the piccolo is much the same as that of the flute, except that it is shorter, being only about half the length of the flute. The fundamental frequency is therefore an octave higher than the flute, its range being from D_5 to B_7. The method of evoking its tones is essentially the same as in the case of the flute. Figure 13–15 shows the waveform of two piccolo notes. The instrument is used to produce the highest notes of the woodwind section in both band and orchestral compositions.

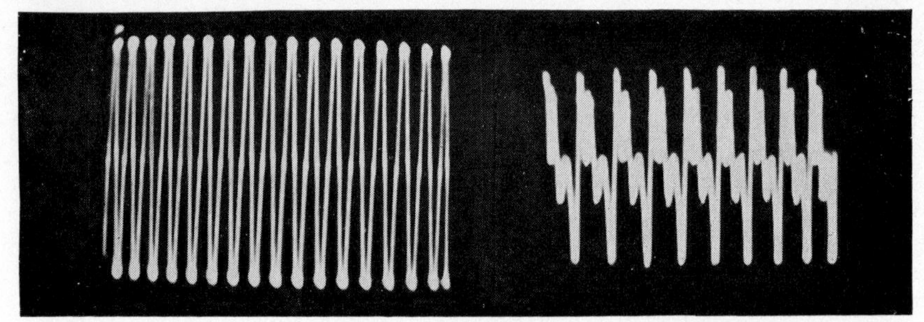

FIG. 13–15. Waveforms of two piccolo notes: *left*, G_6; *right*, G_5.

13–7. *The Clarinet*

We are next to consider the members of a group of wind instruments that make use of a vibrating reed as the sound generator. Like the flute, reed instruments are of ancient origin. Such instruments undoubtedly were used in early Greek times, though they probably did not originate in Greece. The instrument which we know as the **clarinet** appears to have evolved from an instrument known as the **chalumeau** of the middle ages. The records indicate that it was first used in 1751 by Rameau as an orchestral unit.

The modern clarinet is a single-reed type of instrument, in contradistinction to the oboe and bassoon, which have two reeds. Structurally, the clarinet (Fig. 13–16) consists of a cylindrical tube of uniform bore, except for a slight flare near the open end. The tube, which is made of wood or metal, is pierced by 13 to 22 side holes. A system of keys provides for the opening and closing of the holes beyond the six commonly controlled by the finger tips. The mouthpiece, which carries a rather stiff bamboo reed of the beating type, is tapered toward the end. When at rest the end of the reed does not quite touch the slightly beveled embouchure.

As in the case of the reed organ pipe the clarinet has a

FIG. 13–16. Soprano clarinet. (Carl Fischer Musical Instrument Co.)

node at the reed end, and hence would be expected to yield only odd harmonics. Actually, however, while the second harmonic may be absent we find that other even partials are often present. The clarinet, therefore, is not a true closed pipe. When overblown, it sounds the third partial; that is, the pitch goes up a twelfth. The lowest note on the clarinet is D in the bass, and by simple fingering (opening of successive holes) the player can reach B♭ in the treble clef. This range constitutes what might be called the fundamental scale of the instrument. To secure higher frequencies, the natural third harmonic is utilized. To facilitate the formation of three loops in the resonant air column and hence the production of the third partial, a "speaker" key is opened. This key controls a small lateral opening located about six inches from the tip of the mouthpiece. This small vent hole prevents the pressure within the tube from rising above or falling below that of the atmosphere, and thus "encourages" the formation of an antinode in that region. By uncovering this opening, the player is enabled to evoke the third harmonic and thus to shift all of the fundamental tones a twelfth higher. This register, if it may be so called, then begins with B_4 and extends to G_6, this being the natural upper limit of the instrument. However, by utilizing a playing technique known as "cross fingering" a skillful player is able to add another octave to the range. The key system now in general use was introduced by Boehm in 1843. Owing to the difficulties encountered in fingering in certain keys, the clarinet is made in four pitches, viz., C, B♭, A, and E♭, the three latter being what are known as "transposing" instruments. The B♭ instrument sounds a whole tone lower than the written music and the A clarinet three halftones lower. This arrangement is necessary because of the complicated fingering of the instrument.

The clarinet tone has the "reedy" characteristic timbre of the family to which it belongs. The quality of the tone in the lower ("chalumeau") register is decidedly different than that in the higher range. This is evident from an examination of the oscillograms appearing in Fig. 13–17. The reed when sounded alone yields a raucous medley of high frequencies, but when associated with the column of air within the instrument, its vibrations are, in this case, dominated by the resonator, particularly if the reed is not too stiff.

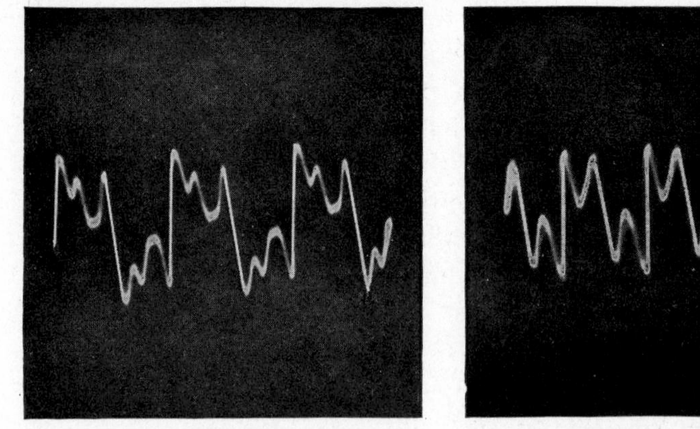

FIG. 13–17. Waveforms of clarinet tones: *left*, G_3; *right*, D_4.

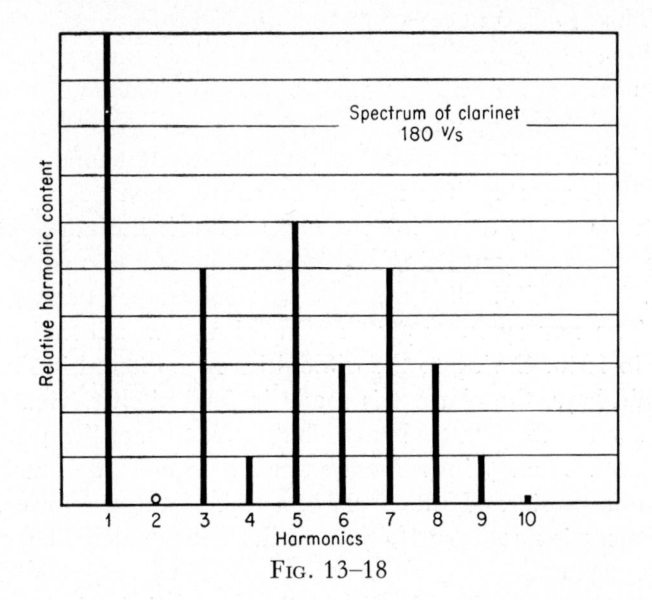

FIG. 13–18

The reed is then compelled to yield notes corresponding to those which are natural to those of the air column. The result of this is that the instrument emits tones which are considerably lower, and much more stable, than those natural to the generator itself. The sound spectrum of a lower note (Fig. 13–18) shows the presence of strong upper partials, the second, fourth, and sixth being particularly prominent. These overtones give the clarinet its characteristic tone color. It is possible to control the intensity of the clarinet tone more effectively than that of other wind instruments.

When used in military and concert bands, the clarinets take the place of the violins in orchestral instrumentation. Indeed, symphonic music is sometimes rendered by largely substituting reeds for strings.

The clarinet has been used extensively in operatic and symphonic renditions since about 1770, one of the themes frequently being carried by this instrument. Wagner gives a prominent part to the A clarinet in the *Tannhäuser* overture. Mozart makes effective use of the clarinet in his *Quintet in A major*.

In addition to the instrument just described and sometimes referred to as the soprano clarinet, a **bass clarinet** is frequently used in orchestral work. As is to be seen in Fig. 13–19, it resembles somewhat the saxophone in appearance, being crooked at the upper part and having a turned-up metal bell. This instrument is made in both A and B$\flat$ types, though the latter form is more commonly used. The bass clarinet is tuned an octave below the corresponding soprano instrument, and has a range extending from D_2 to $E\flat_5$. Having this range it can compass the higher notes of the bassoon and the lower notes of the flute. Because of its musical flexibility and rich tonal quality, it is used by many composers to carry solo and theme parts. Wagner for instance uses it for important passages in several of his compositions, notably in *Tannhäuser* and in *Die Valkyrie*.

In passing, it is interesting to note that some musicians contend that an A clarinet, for instance, will yield a different tone color than that produced by a B$\flat$ instrument. Objective tests show that few, if any, composers or conductors can actually detect any

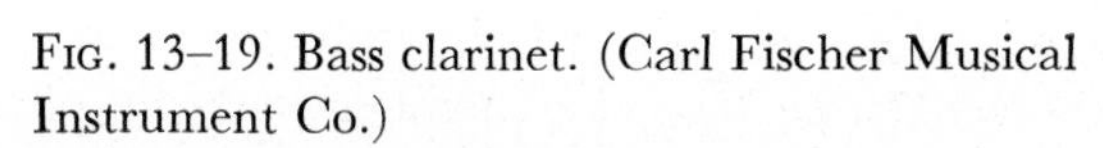

FIG. 13–19. Bass clarinet. (Carl Fischer Musical Instrument Co.)

difference in sound quality between the A and B♭ clarinets, and the same is probably true concerning the A and B♭ trumpets. Any apparent difference in tone color is probably subjective.

Readers who are interested in the technique of tone color control in clarinet playing may wish to read an article appearing in the July, 1948, issue of *The Etude*, entitled The Foundation of Clarinet Tone Quality by Paul Van Bodegraven.

13–8. *The Saxophone*

Because of its wide use in dance bands it is popularly supposed that the saxophone is of quite recent origin. As a matter of fact, it was invented by Adolphe Sax, a skilled Belgian instrument maker, in 1840 and made its first appearance in a symphony orchestra in 1844. It is a single reed organization, but has a conical tube, and the diameter of the tube at the reed end is larger than that of other reed instruments (Fig. 13–20). This means that the acoustical coupling between the reed and the pipe is not so intimate as in the case of the clarinet. It may be shown by theoretical analysis that a conical tube, acoustically closed at the generator end, will give rise to a complete complement of overtones, and these mathematical deductions are realized in practice. Further, the harmonics developed in a conical air column are found to be the same as those which obtain in the case of a cylindrical column open at both ends, and of the same length. In the case of the saxophone, because of the relatively thick reed commonly employed with this instrument, there is more or less conflict between this vibratile member and the air

Fig. 13–20. Tenor saxophone. (C. G. Conn, Ltd.)

column as to which vibrating body shall control the pitch of the note emitted. This "struggle" between the reed and the air column gives rise to a timbre which combines some of the characteristics of the clarinet and the cello. In order to reduce the impedance of the

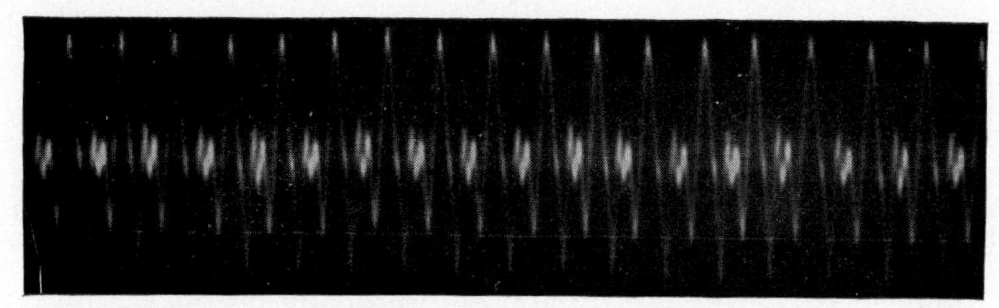

FIG. 13–21. Waveform of the saxophone in the middle register.

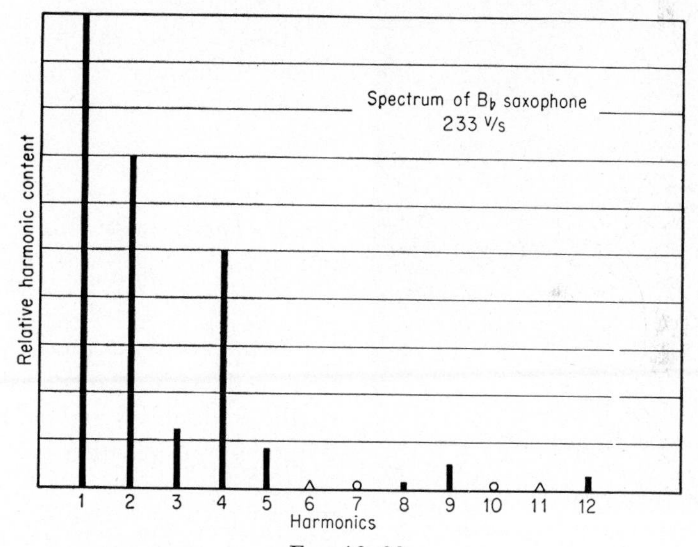

FIG. 13–22

air column, and therefore permit it to control the pitch, the holes are made relatively large; and this necessitates the use of keys for all openings. In this country the saxophone is widely used in dance bands, and occasionally in orchestras. A typical saxophone waveform is shown in Fig. 13–21 and the corresponding spectrum in Fig. 13–22. This spectrum is more or less unique. It will be observed that while a small percentage of the ninth and twelfth harmonics is present several of the adjacent partials are either absent or represented only by a trace.

The several types of saxophones have ranges as follows: soprano, $A\flat_3$ to $E\flat_6$; alto $D\flat_3$ to $A\flat_5$; tenor, $A\flat_2$ to $E\flat_5$; baritone, $D\flat_2$ to $A\flat_4$; base, $A\flat_1$ to $D\flat_4$.

As indicated above, the saxophone was originally employed in the rendition of serious music, one of the first instances being in a composition by Bizet entitled *L'Arlésienne Suite*. The saxophone is assigned a conspicuous voice by Ravel in his well-known "Bolero." Most brass bands now employ several saxophones; its use in dance orchestras has already been noted. Perhaps no other musical instrument has been so widely abused, musically, as the saxophone, but in the hands of skilled performers it is capable of yielding useful musical sounds.

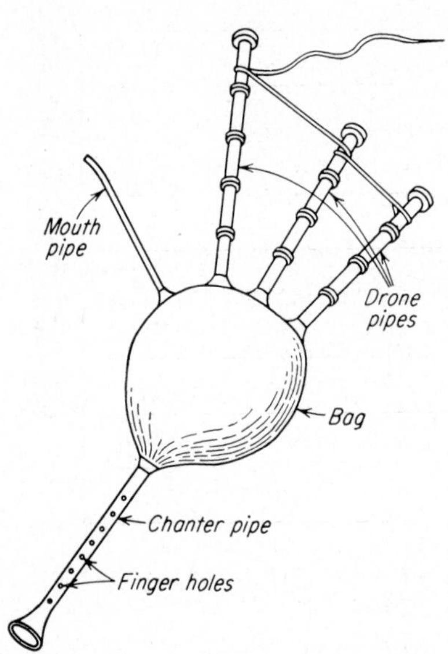

FIG. 13–23. Sketch of a bagpipe. (After H. F. Olson, "Musical Engineering," McGraw-Hill Book Company, Inc.)

Though the saxophone is made of metal it is commonly classified as a wood-wind instrument.

13–9. *The Bagpipe*

Perhaps the most unusual reed instrument is the bagpipe. It consists of four pipes each equipped with a reed, of the single or double type, supplied with air from a leather bag kept inflated by the player's breath. Three of the pipes, as indicated in Fig. 13–23, are called drones; these sound continuously.

One of the pipes, called the chanter, is pierced with eight holes that can be closed with the fingers, thus making it possible to vary the pitch of the sound emitted by this particular pipe. The melody is played by means of the chanter, and the drones furnish a continuous harmonic background of sound.

13–10. *The Oboe*

In the oboe we have an example of an instrument in which the generator consists of two beating reeds, as compared to one in the clarinet. The oboe also differs in another important particular from the clarinet; it has a conical instead of a cylindrical tube, the double reed being positioned at the apex of the cone. Figure 13–24 shows the general features of the instrument.

The reeds of the oboe are of cane, but they are much smaller and lighter than the type used with the clarinet. The two reeds of the oboe when at rest form a small elliptically shaped opening at their free ends. Under the action of the player's breath the reeds vibrate transversely and longitudinally. This motion alternately opens and closes the opening, thus giving rise to pulses of compression within the tube. To an even greater extent than in the clarinet, the resonant air column controls the frequency of the vibration of the double-reed system. The fingering technique is theoretically similar to that of the flute—a pipe open at both ends. Because of the fact that a delicate adjustment is required in order to effect a change of pitch it is customary to have the oboe sound the pitch (A) for the orchestra as a whole. The waveform of a representative oboe tone is shown in Fig. 13–25. An analysis of the characteristic sound of the oboe (Fig. 13–26) shows a number of both even and odd harmonics to be present, as is to be expected, the octave being stronger than the

FIG. 13–24. Oboe. (Carl Fischer Musical Instrument Co.)

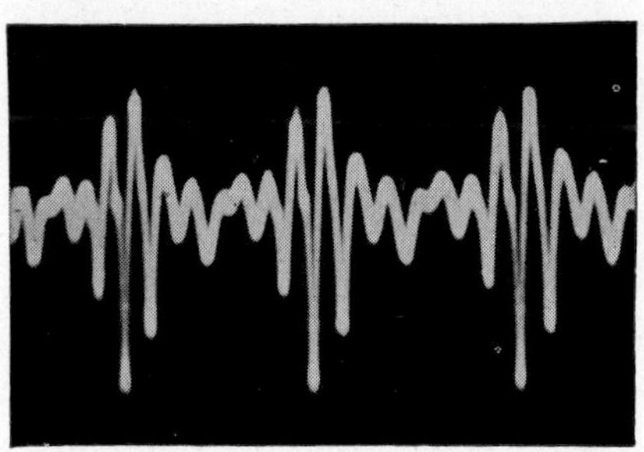

FIG. 13–25. Waveform of oboe sounding C_5.

fundamental; the third and fifth harmonics are also relatively strong. This complement of harmonics gives the oboe its characteristic tone color, a penetrating, "reedy" sound resembling in some respects the human voice. Overblowing gives the octave rather than the third partial, as in the case of the clarinet. This fact simplifies somewhat the note-hole arrangement as compared with that of the clarinet.

The range of the oboe is from B$\flat_3$ to F_6 inclusive.

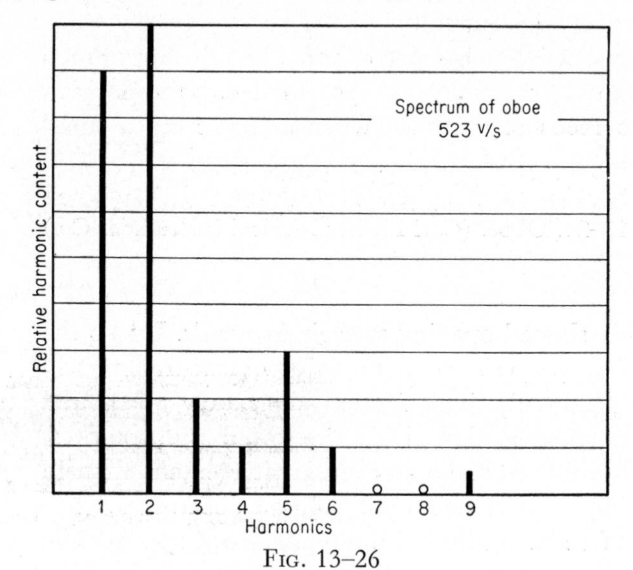

Fig. 13–26

The oboe has long been used effectively as an orchestral instrument—from the time of Bach and Handel down to the works of Sibelius. A beautiful oboe obbligato appears in the chorale *Jesu, Joy of Man's Desiring*. A notable solo passage for the oboe occurs in the first movement of Beethoven's *Fifth Symphony*.

13–11. *The English Horn*

This instrument, a picture of which is shown in Fig. 13–27, is the alto member of the double-reed family. While somewhat similar in appearance to the oboe, this reed instrument has a curved mouthpipe, is somewhat larger than the oboe, and terminates in a globular-shaped bell. Acoustically, the English horn develops both odd- and even-numbered partials, but its complement of overtones is somewhat different from the oboe, as may be seen by comparing the sound spectra shown in Figs. 13–26 and 13–29. Its waveform is indicated in Fig. 13–28. The English horn is pitched a fifth lower than the oboe, and like most lower pitched instruments has a relatively extensive complement of upper partials, as shown in Fig. 13–29. Its tone is, accordingly, richer and somewhat more somber than that of the oboe, the tone character resembling, to some extent, the tone of the human voice. The instrument is sometimes used in orchestral compositions to express grief and anguish, sometimes tenderness or a dreamy mood. The English horn is built in the key of F, and has a pitch range of two and a half octaves, its lowest note being E below middle C. The keying arrangement and the fingering are the same as those of the oboe. In its modern form, the English horn was first

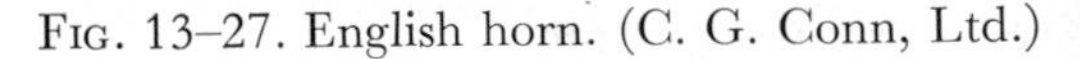

FIG. 13–27. English horn. (C. G. Conn, Ltd.)

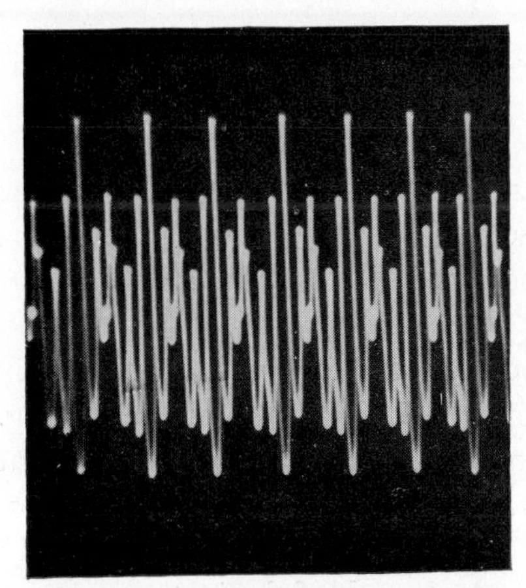

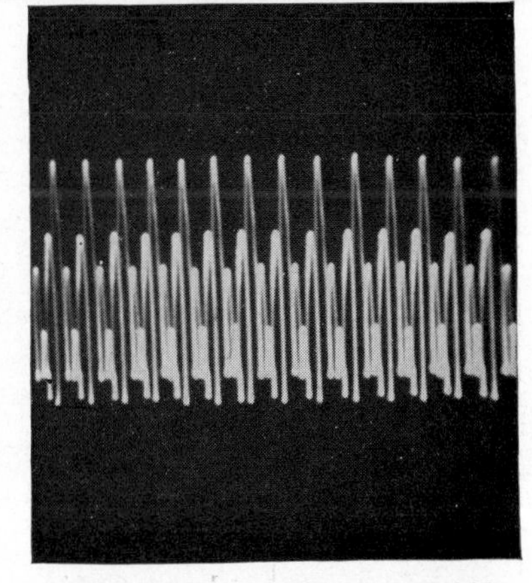

FIG. 13–28. Waveforms of English horn, sounding A 220 and A 440.

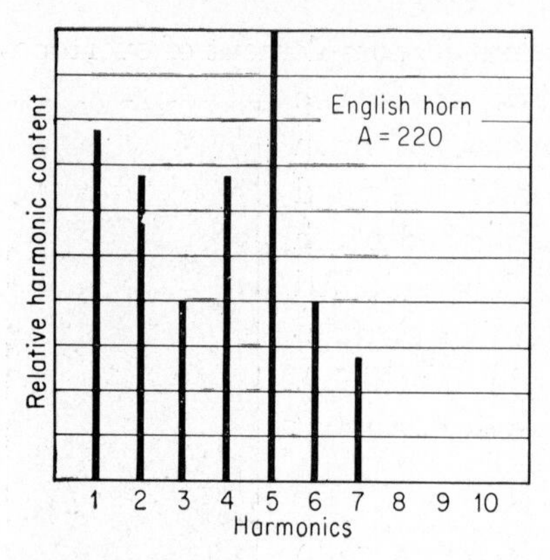

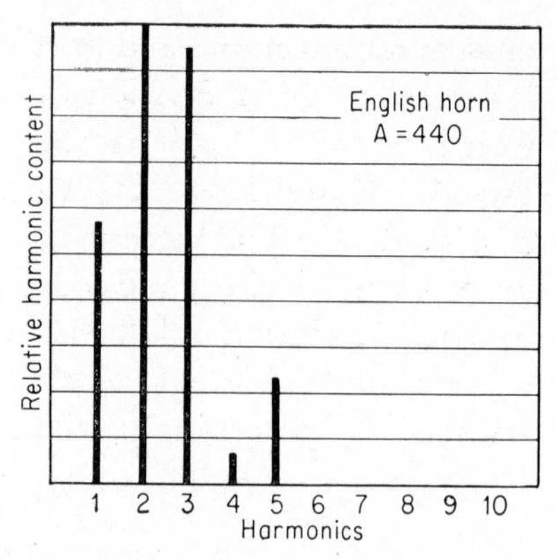

FIG. 13–29

used by Rossini in *William Tell* (1829) and by Meyerbeer in *Roberto* in 1831. Commonly only one English horn is used in orchestral instrumentation.

13–12. *The Bassoon*

Like the oboe, the bassoon is a double-reed instrument, and it functions, essentially, as an alto oboe. It consists of a conical tube 93 in. in length, measuring 1¾ in. in diameter at the flared end, and 3⁄16 in. at the reed end. A short metal crook connects the reed assembly to the wooded tube. The double reed is larger than that used on the oboe. Because of its relatively great length it is doubled back on itself (Fig. 13–30), the actual length of the instrument being only 4 ft. The acoustical length of the bassoon is controlled by seven holes and 16 or 17 keys. Owing to the length of the instrument the

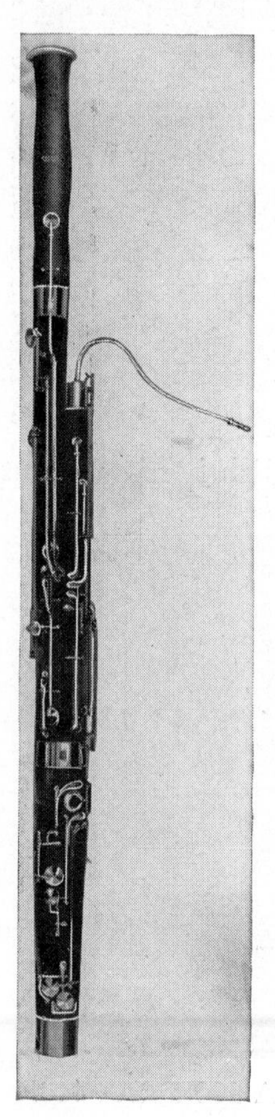

FIG. 13–30. Bassoon. (Carl Fischer Musical Instrument Co.)

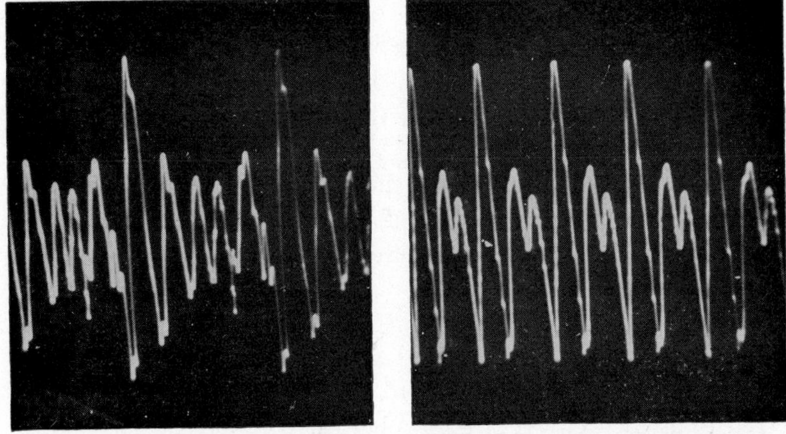

Fig. 13–31. Waveforms of the bassoon in the low and middle register.

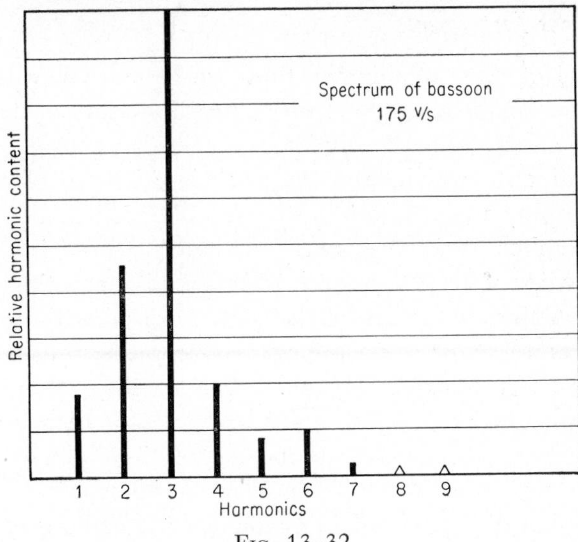

Fig. 13–32

holes pass obliquely through the tube wall, thus bringing the actual apertures within the compass of the fingers. Representative waveforms emitted by the bassoon are shown in Fig. 13–31, and in Fig. 13–32 may be seen the spectrum of the middle register. The instrument yields the full harmonic series and has a range of more than three octaves, extending from $B\flat_1$ to $E\flat_5$. It will be noted that the strongest component is the third harmonic, the fundamental being relatively weak. The three registers of the bassoon, like those of the clarinet, each have distinct tonal characteristics. The bassoon has been long and extensively used in orchestral renditions. Notwithstanding the fact that the construction of this instrument violates some acoustical principles, its tonal color and its mechanical design are such that it gives wide scope to the player, being sometimes used for solo melodies. For instance, Beethoven used it extensively, having written for it a number of independent parts. Grieg makes effective use of the bassoon in "In the Hall of the Mountain King" in his *Peer Gynt Suite;* Weber in his *Hungarian Fantasia* gives it a solo passage. When played staccato, the instrument produces a more or less humorous effect. For this reason the bassoon is sometimes referred to as the "clown of the orchestra."

13–13. *Brass Instruments*

In several of the preceding sections we have dealt with instruments having single or double reeds that function as the sound-generating component. We are now to consider another group of wind instruments which, though not classified as of the reed type, do employ what amounts to a double reed as the primary vibratile member. Reference is here made to the brass instruments, in which the lips of the player serve as the sound-generating element. In these instruments the lips of the player are drawn more or less tightly across a conical or cup-shaped mouthpiece, thus producing a double membranous reed. The breath of the player directed against these two bands of stretched tissue causes them to vibrate at a frequency which depends upon their tension, effective length, and the air pressure. We have here an instance of an outward striking reed as distinguished from the inward striking type used, for instance, in such instruments as the oboe.

The resonant tubes with which these mouthpieces are associated are of various forms. In general, however, the tubes are conical in shape, for at least a part of their length, and end in a more or less broadly flaring bell. In most instruments of this type the length of the tube is great compared to its average diameter. The shape of the tube, particularly in the enlarged section, exerts a pronounced influence on the tone color of such instruments. One effect of the terminal flare is to reduce the intensity of the higher partials. A horn with a wide spreading bell produces a tone that might be characterized as "diffused"; while a small bell yields a "brighter" and a more penetrating quality. Another effect of the enlarged section of the tube involves the effectiveness of the horn as a radiator of sound. The gradually increasing curvature tends to cause the sound disturbance to emerge in the form of spherical waves. In consequence of this, the energy is transferred to the ambient air more efficiently.

A horn having a properly designed terminal section functions as an open pipe, though actually it is closed by the player's lips at the mouthpiece end. It yields a full complement of upper partials. In general, the fundamental tone of a brass instrument is produced only with considerable difficulty.

It is to be noted that the shape of the mouthpiece produces a decided effect on the tone quality produced by brass wind instruments. A deep conical mouthpiece assists in the production of a smooth tone; while a shallow cup-shaped mouthpiece facilitates the formation of the higher partials, and hence a "bright" tone. Each type of horn requires a particular form of mouthpiece. Two representative mouthpiece units are shown in Fig. 13–33.

Referring again to the shape of the longitudinal section of a brass instrument, it is interesting to note that there does not appear to be any unanimity of opinion as to the optimal *flare coefficient* in horn design. By this term is meant the rate of increase in diameter with distance along the tube axis. Dr. E. G. Richardson in his well-known work entitled "The Acoustics of Orchestral Instruments and the Organ" holds that the taper of brass instruments follows an exponential law. American designers do not concur in this opinion. For example, Mr. Vincent Bach, who is a skilled trumpet player and a manufacturer of high-grade brass instruments, does not subscribe to the exponential idea. From what the author can learn, it would appear that each designer of such instruments has his own flare coefficient, and that the value of this factor has been arrived at largely on an experimental basis, rather than from theoretical considerations. Here we find a situation similar to that encountered in the design and construction of the violin.

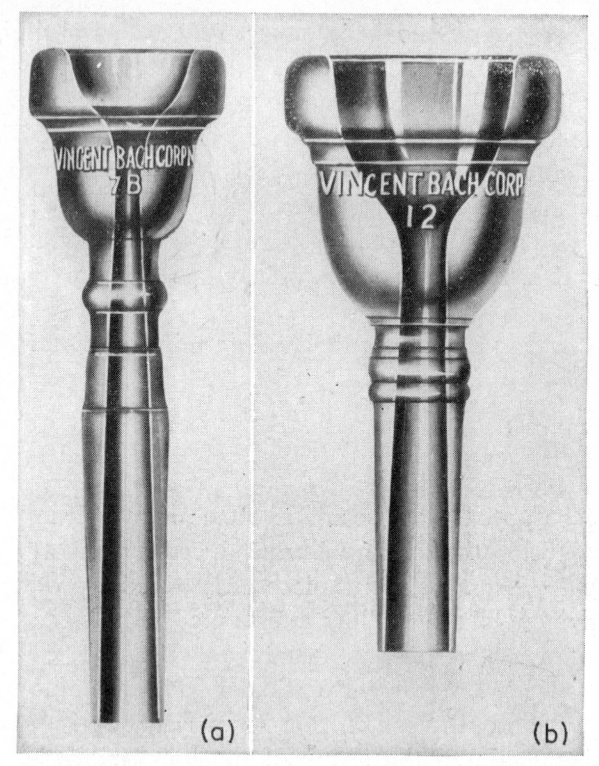

FIG. 13–33. Horn mouthpieces: (*a*) trumpet; (*b*) trombone. (Vincent Bach Corporation.)

In succeeding sections we shall consider the design and functioning of several representative instruments of the brass choir.

13–14. *The Horn*

The simplest form of horn is that sometimes referred to as the "natural" horn, an illustration of which appears as Fig. 13–34. It consists of a tube some 10 ft in length and is equipped with a funnel-shaped mouthpiece. In the original form of this instrument no provision was made for altering the length of the resonant air column. The player, therefore, could produce the harmonic scale by means of the lips alone. However, by introducing the hand, fingers first, into the bell of the horn in such a manner as not to completely close the opening, it is possible to lower the fundamental a semitone and thus establish a new series of partials. Thus certain additional notes were added. Though used by the older masters, it has largely been replaced in the modern orchestra by an instrument in which provision is made for changing the length of the resonant air column. This is accomplished by means of a system of pistons and crooks.

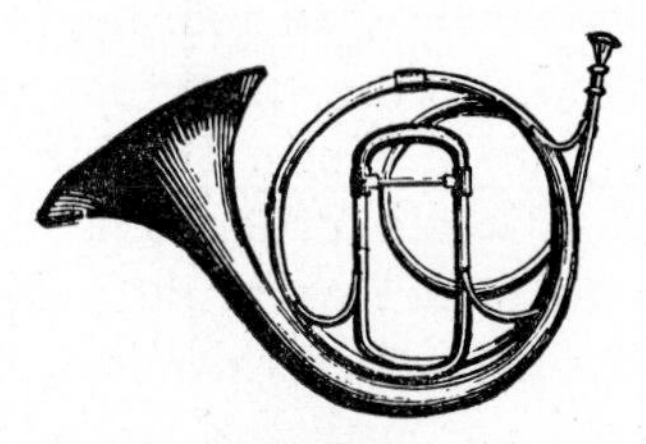

Fig. 13–34. Original form of horn.

The **valve horn,** commonly known as the **French horn,** is the modern counterpart of the original natural horn; it has a tube length of from 12 to 16 ft. A cut of this instrument is shown in Fig. 13–35. By means of three valves (of the rotary type) one or more side-tubes, or "crooks," as they are called, are added to the original length of the tube, thus augmenting the length of the resonating air column, and thereby lowering the pitch by one, two, or three semitones when used singly. When the valves are

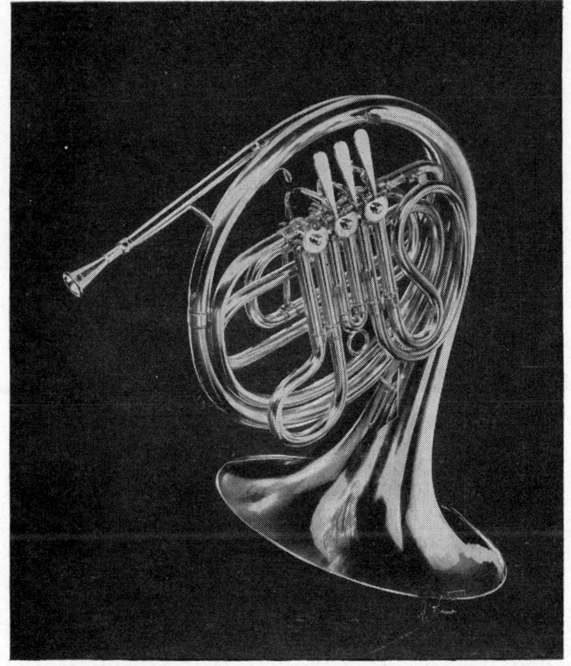

Fig. 13–35. Modern French horn. (C. G. Conn, Ltd.)

not depressed by the player, the side-tubes are closed; the valves may be depressed singly or simultaneously. This valve mechanism thus makes possible the full chromatic scale, and thereby gives to the brass instrument a facility of execution comparable to that of the wood winds. The introduction of the hand into the bell is also practiced in order to facilitate the production of certain notes, and also to act as a mute. However, in certain modern high-grade horns an extra valve is provided for muted tones. A careful analysis of the acoustical situation will reveal the fact that, since any given valve has the effect of opening the tube and introducing an additional length of tubing, the simultaneous use of two valves will

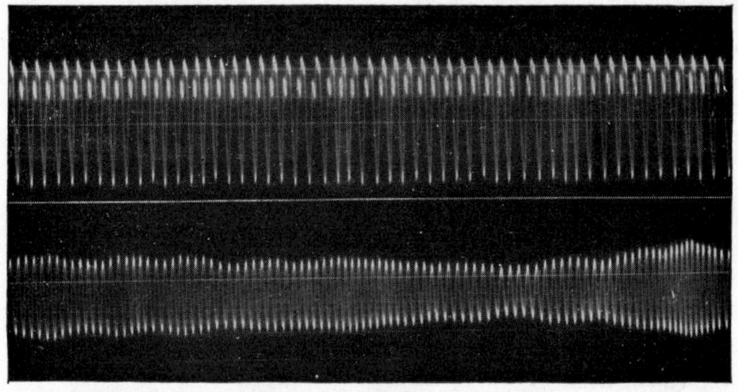

FIG. 13–36. Waveforms of French horn in middle and high register.

lead to imperfect intonation unless some compensating device is available. This limitation is overcome, in the case of some horns at least, by an arrangement whereby short compensating tubes are automatically introduced when two or more pistons are simultaneously depressed. The horn is made in several keys, but the F horn, owing to its convenience of playing in different keys, is generally used.

The range of the French horn extends from B_1 to F_5. Its best range is from the fourth to the twelfth harmonic. The waveforms of two different notes sounded on this instrument are shown in Fig. 13–36. In Fig. 13–37 we see the spectrum of the tone having the lower pitch shown in Fig. 13–36. It will be observed that in the case of the lower note the second harmonic is louder than the fundamental, which is relatively weak, and that the third harmonic is also a strong component. In the region of F_2 a large number of both odd and even upper partials appear. However in the higher range, even when blown fortissimo, the instrument yields a tone that shows almost no harmonic content other than the fundamental.

The tone of the French horn is characterized by a soft plaintive quality, which makes it blend well with both the wood winds and the brass instruments. In orchestral work the instrument is frequently used to fill in and sustain the principal harmony; in solo passages it is assigned melodies of a sustained nature. Frequently French horns are used in pairs, half the players taking the lower notes and the others the higher. This is because players find it difficult to adjust the lips readily to cover the entire scale. In the modern orchestra, however, the players are usually expected to be able to compass the entire scale. In Mendelssohn's "Nocturne" from *A Midsummer-Night's Dream* is to be found one of the most beautiful passages for horns in all musical literature. A short but very striking passage for the French horn appears in "The Merry Pranks of Til Eulenspiegel" by Richard Strauss.

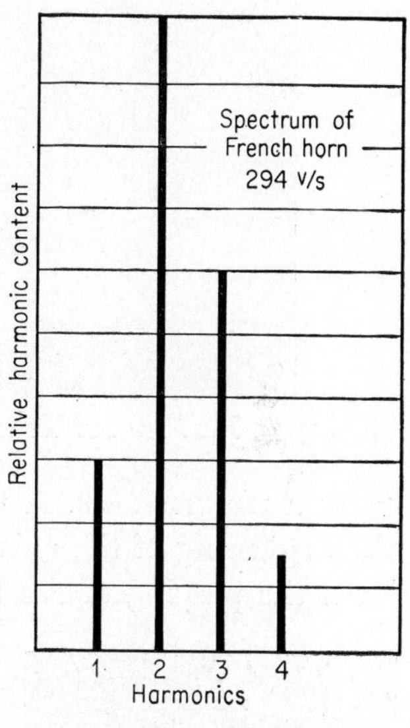

FIG. 13–37

13–15. *The Trumpet*

Another brass instrument that is widely used in orchestral work is the trumpet. From the illustration (Fig. 13–38) it will be seen that the total tube length is less than that of the horn, and is cylindrical throughout approximately three-fourths of its length. From there it widens conically, ending in a flare at the bell, whose terminal diameter is about 4 in. The diameter of the cylindrical part of the tube is only ⅜ in. The resulting small diameter-to-length ratio, together with the use of a cupped mouthpiece, makes possible the production of a full series of harmonics up to and including the tenth.

In the trumpet, and certain other brass instruments, each of the valves instead of being of the rotary type, as in the case of the horn, consists of a plunger-like mechanism which is referred to as a piston valve. A tuning slide is provided in the bend near the beginning of the enlarged portion of the tube.

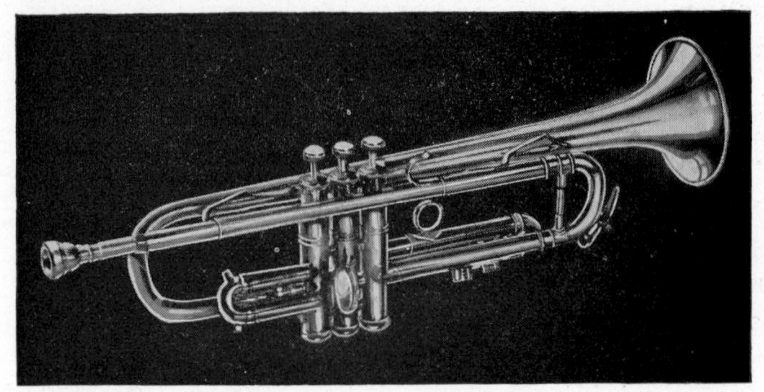

Fig. 13–38. Trumpet. (Vincent Bach Corporation.)

In the playing of the trumpet and other brass instruments, one encounters the problem of attaining correct intonation, particulary when using the valves. By the use of a properly designed and fitted mouthpiece, a skilled player can largely compensate for any inherent intonation defects. The interested reader is referred to a paper appearing in the September, 1950, issue of *Symphony* entitled Problems in Intonation of Brass Instruments, by Vincent Bach. In addition to the subject of intonation, Mr. Bach touches upon another important, and related, point. He takes the position that symphony orchestras should tune the string section to A by means of a bar or other device, while the brass choir should be tuned to a B♭ tuning standard, thus making it possible for them to tune to an open tone. The author concurs in Mr. Bach's opinion regarding such a tuning procedure.

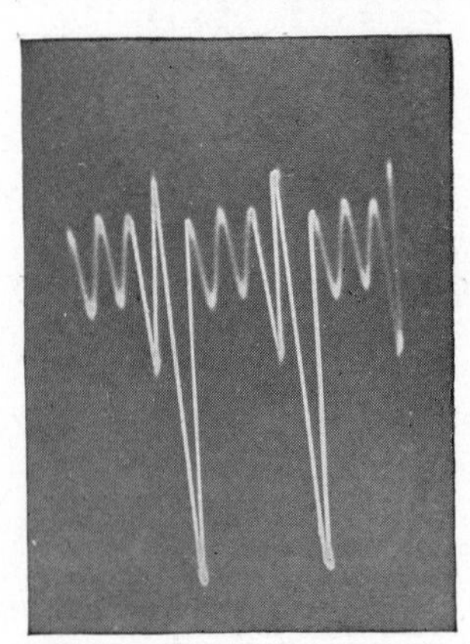

Fig. 13–39. Waveform of trumpet when sounding F_4.

In order to facilitate fingering, the trumpet is commonly designed for use in two pitches, B♭ and A, though an E♭ instrument is sometimes used in military bands. This means, as before indicated, that when C is written in ordinary music, B♭ is sounded by the player. The trumpet is therefore a transposing instrument.

An oscillogram showing a characteristic waveform of the trumpet appears in Fig. 13–39, and the corresponding spectrum is shown in Fig. 13–40. These records show that the first four harmonics are all strong, the second being as strong as the fundamental, and

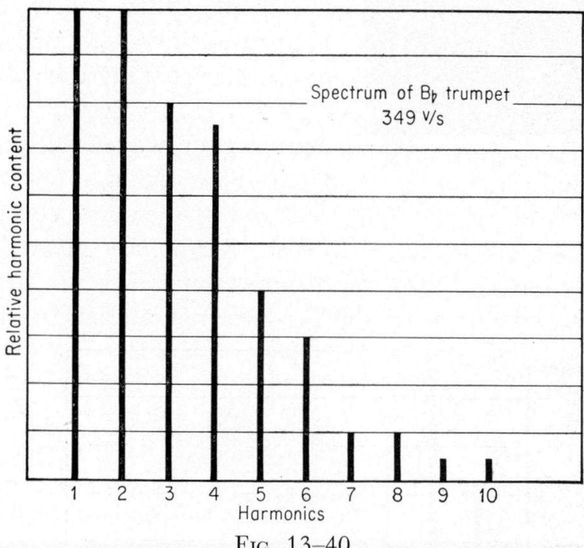

FIG. 13–40

that the remaining partials diminish in magnitude as the order increases. It is this spectral distribution of energy which gives to the trumpet its full and penetrating tonal characteristics. The range of the trumpet extends from E_3 to $B\flat_5$, thus giving it a compass of two and one-half octaves. The trumpet is the soprano instrument of the brass orchestral choir.

In the "Marche Slave" and the "1812 Overture" Tschaikowsky has assigned passages of great beauty to the trumpets.

13-16. *The Cornet*

In design the cornet differs from the trumpet in that its tube is shorter and the bore is larger, a greater percentage being conical.

As a consequence of the larger bore, the harmonics above the seventh cannot readily be produced. Because of the smaller number of partials evoked when blowing the cornet, this instrument does not require so careful an adjustment of the lip tension and air pressure as is required when sounding the trumpet. The cornet is therefore less difficult to play than the trumpet. In fact, the facility with which it can be played probably exceeds that of all other brass instruments. The characteristic waveform of the cornet is shown in Fig. 13–41, and a representative sound spectrum in Fig. 13–42. It will be noted that the harmonic content is by no means the same as that of the trumpet; in the cornet the second harmonic is stronger than the fundamental and there are fewer

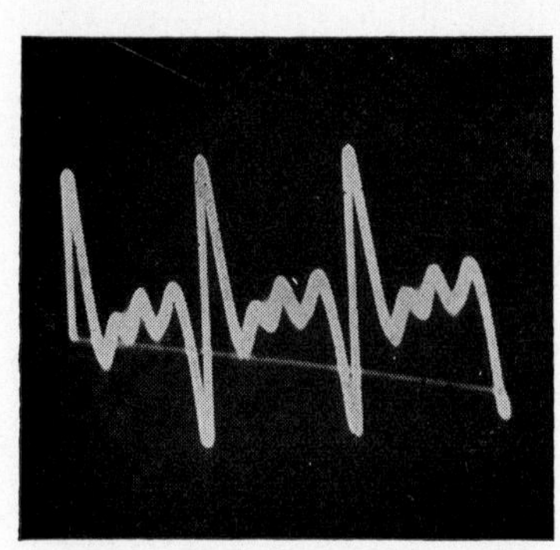

FIG. 13–41. Waveform of B♭ cornet, sounding A 440.

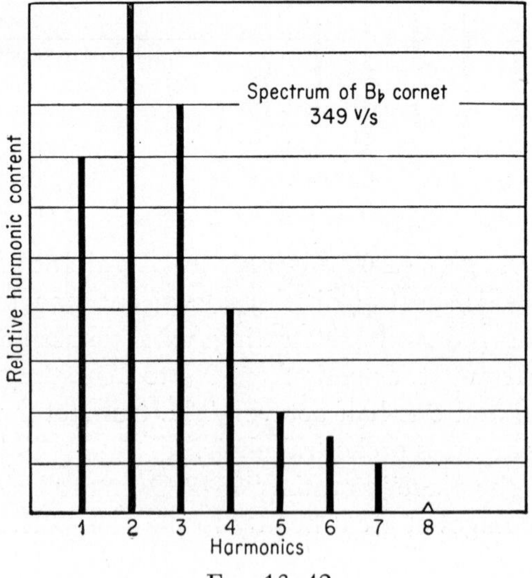

FIG. 13–42

partials than in the trumpet. The strong first and second upper partials give to the cornet its characteristic timbre—a brassy piercing tone. The range of the cornet is approximately the same as that of the trumpet, and it is likewise used in the B♭ and A types.

The cornet is used principally in military bands, frequently carrying the lead part; occasionally it is used in symphonic renditions, as in César Franck's *Symphony in D Minor*.

FIG. 13–43. Tuba. (C. G. Conn, Ltd.)

13–17. *The Tuba*

The tuba, the fourth member of the brass choir, serves as the bass of the brass group. This instrument has a large conical bore throughout most of its length (about 18 ft), and terminates in a wide bell, one form of the instrument being shown in Fig. 13–43. The waveform of a note in its midrange is seen in Fig. 13–44, and the corresponding spectrum in Fig. 13–45. The tuba is commonly equipped with three valves, but in some of the newer models a fourth valve is added. The instrument is made in two types—the E♭ and the B♭. The range of the E♭ tuba covers about two and a

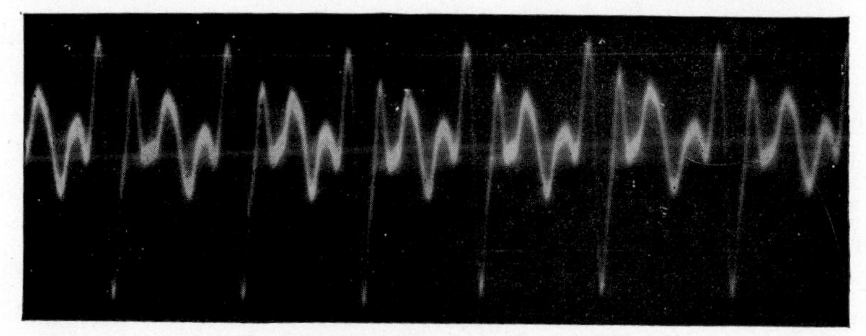

FIG. 13–44. Waveform of tuba.

half octaves, from A_1 to $E\flat_4$. The range of the BB♭ instrument is from E_1 to $B\flat_3$. Owing to the wide bore of the tube, its fundamental is readily evoked by the player. It will be seen that the sound spectrum of the tuba is characterized by a limited number of harmonics of which the fundamental and the octave are the principal components. This and its low pitch account for its sonorous tone.

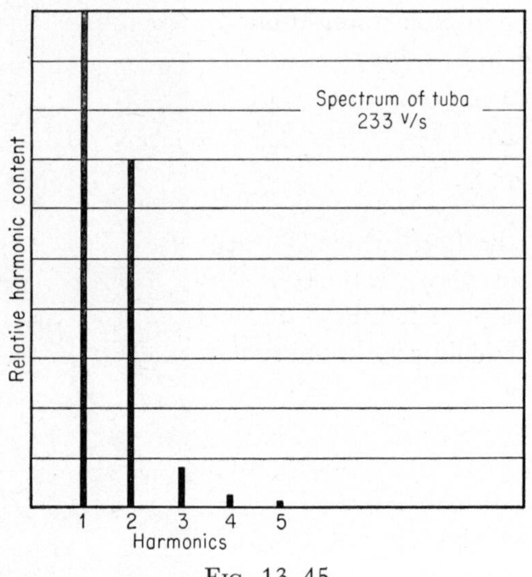

FIG. 13–45

Dvořák has used the tuba effectively in the *New World Symphony*. Another good example is to be found in *Finlandia*, by Sibelius.

13–18. *The Trombone*

The unique instrument among all of the wood winds and the brasses is the trombone (Fig. 13–46). Here we have an instrument without pistons, yet one which is capable of yielding the chromatic scale throughout more than three octaves. Throughout the major portion of its length it consists of a telescoping cylindrical tube terminating in a bell of moderate size; a cup-shaped mouthpiece is used. By making use of a U-shaped sliding crook, one side of which articulates with the mouthpiece portion, the acoustical length of the resonant air column can be adjusted within wide limits. In practice seven positions of the slide are commonly utilized. When the slide is closed, we have what is designated as the first position; the fundamental and seven upper partials can then be evoked.

Each successive position lowers the note by a semitone. In the last position the fundamental is not obtained. The pressure of the breath and the tension of the lips also assist in producing a given note. Because of the slide-tube feature it is possible for the performer to play in true intonation. In this respect the trombone is comparable to the violin.

FIG. 13–46. Trombone. (Vincent Bach Corporation.)

A representative waveform of the trombone is shown in Fig. 13–47; the analysis of a characteristic wave appears in Fig. 13–48. Because of the shape of the resonant air column, the timbre of the trombone resembles somewhat that of the trumpet, as may be seen by comparing the sound spectra of these two instruments. It is not, however, as rich in high upper partials as the trumpet. Its tone is, therefore, characterized by a more noble and sonorous

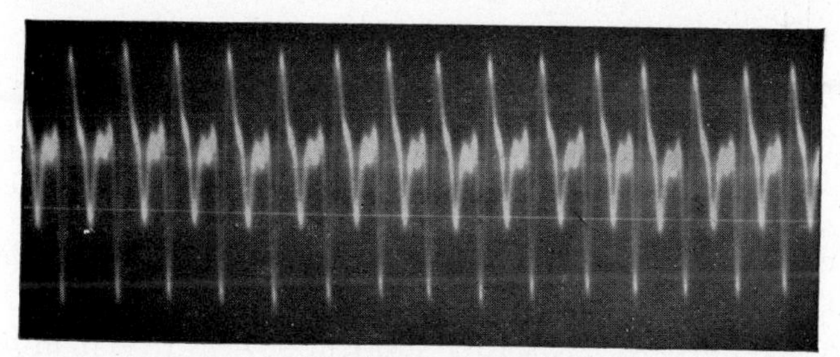

FIG. 13–47. Waveform of trombone.

quality. The instrument has a wide dynamic range. The trombone is made in several sizes, the tenor instrument in B♭ being most commonly used. Its compass is from E_2 to B_4. Because of the noble dignity of its tone the trombone was for a long period in medieval times used in connection with religious services. In rendering symphonic compositions the modern orchestra commonly makes use of three trombones, one of which (the bass trombone) is pitched considerably lower than the tenor instrument. Trombones speak a dramatic passage in the prelude to Wagner's *Lohengrin*. Another well-known example of the use of trombones occurs in the "Triumphal March" from Verdi's *Aïda*.

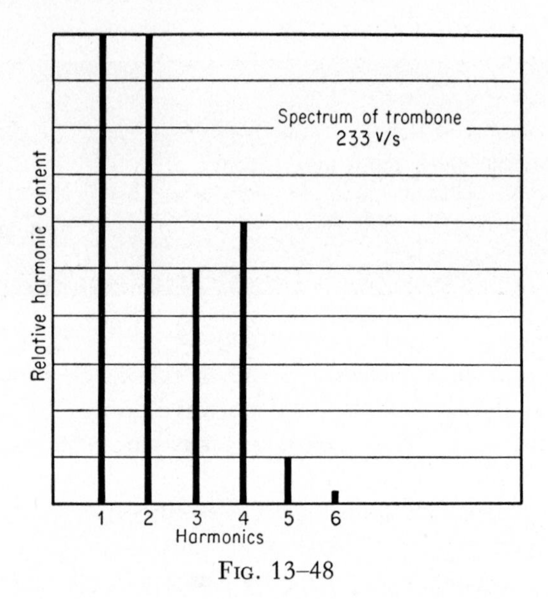

FIG. 13–48

The base trombone, referred to above, is similar to the tenor instrument in design and construction. Its range is from A_1 to $G\flat_2$.

13–19. *The Vocal Organs and the Voice*

The most perfect musical instrument known to man is the human larynx. In possible variations of pitch, timbre, and intensity, it has not been equaled by any of the musical devices that man has yet devised. Essentially the human vocal apparatus is a wind instrument of the double-reed type. The voice is produced by the forcing of air from the lungs through the opening (glottis) between two adjacent pieces of membranous tissue known as the vocal cords (Fig. 13–49), thus causing the free edges to vibrate as in any double-reed instrument. By means of muscles connecting these vibratile tissues with the walls of the larynx, the tension, the length, and the thickness of the vocal cords can be modified. This makes possible not only an adjustment of pitch but also affords some control of the quality. When singing a note, the inner edges of the cords probably touch during each vibration, thus alternately

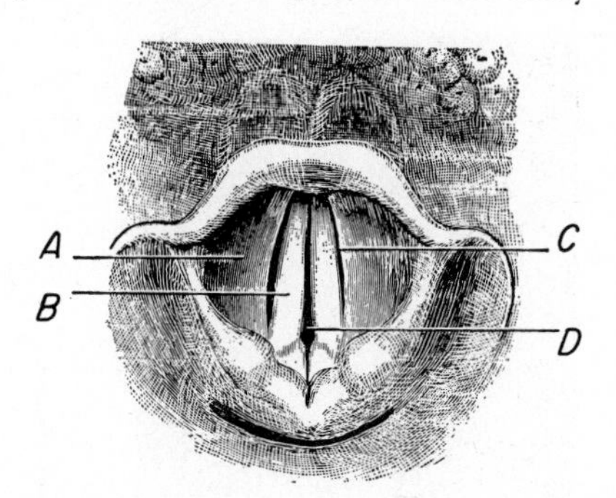

FIG. 13–49. Laryngoscopic view of vocal cords and associated parts when a high note is being sung. *A*, false vocal cords; *B*, true vocal cords; *C*, ventricles; *D*, rima glottidis (opening). (From "Morris' Human Anatomy") by Schaeffer, McGraw-Hill Book Company, Inc.)

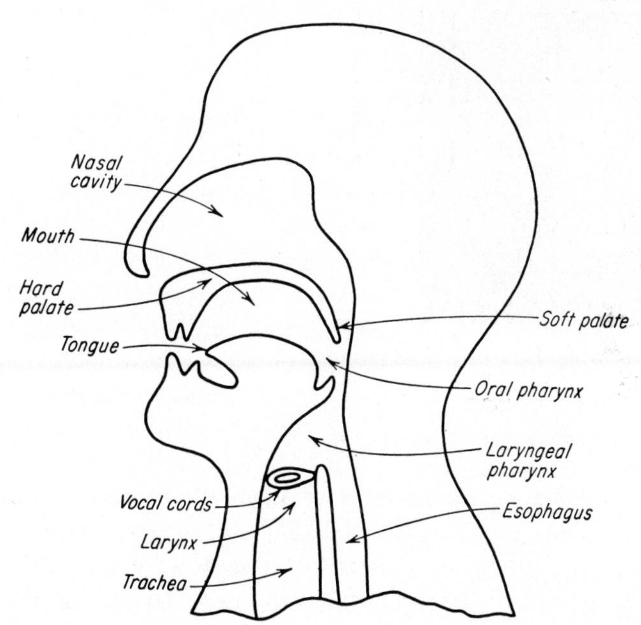

FIG. 13–50. Outline sketch showing mouth and head cavities.

opening and closing the glottis. The resulting puffs of air, under pressure from the lungs, serve to set up resonant vibrations of the air in the associated cavities of the throat, mouth, and nose. The unique feature of the human vocal apparatus is that the size and shape of the resonant system is under the conscious control of the speaker or singer. The oral and nasal cavities, the larynx, and the vocal cords constitute what might be termed an "acoustical network." The diagrammatic sketch appearing as Fig. 13–50 indicates the principal elements composing this network. The resonant frequency of this system of interconnected cavities may be changed by altering the size and contour of the oral cavity. The lower jaw, the tongue, and the lips, singly or acting together, serve to accomplish this end.

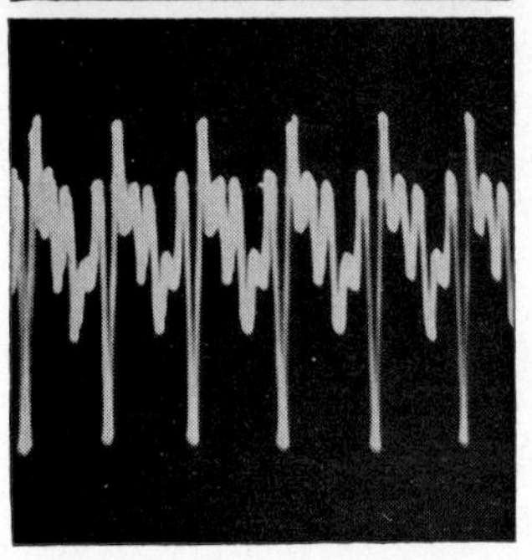

FIG. 13–51. Waveforms of two voices intoning the same vowel at approximately the same frequency.

While the vocal organs constitute an exceedingly flexible acoustical organization, it is to be noted that a given set of vocal cords, associated as they are with a particular resonating system, will produce a perfectly definite and characteristic timbre—the voice of a particular person. Certain harmonics will be emphasized while others will be suppressed. In Fig. 13–51 is shown the waveform of two different female voices as recorded when each intoned the syllable *ah* on the same pitch. The marked difference in waveform is obvious. The waveform of a person's voice is as characteristic of that individual as his fingerprints. An analysis of such voice waveforms shows that, in some cases at least, certain of the upper partials are inharmonic. Incidentally, it may be added that partials up to the thirty-fifth have been noted in vocal sounds. The extent to which the timbre of the voice may be modified by a vocalist in singing a note is strikingly evidenced by the two records shown in Fig. 13–52.

A great amount of research has been carried out in an effort to arrive at an understanding of the acoustical structure of both vowel and consonantal sounds. Such knowledge is highly desirable if one s to make even a beginning in the improvement of the singing or

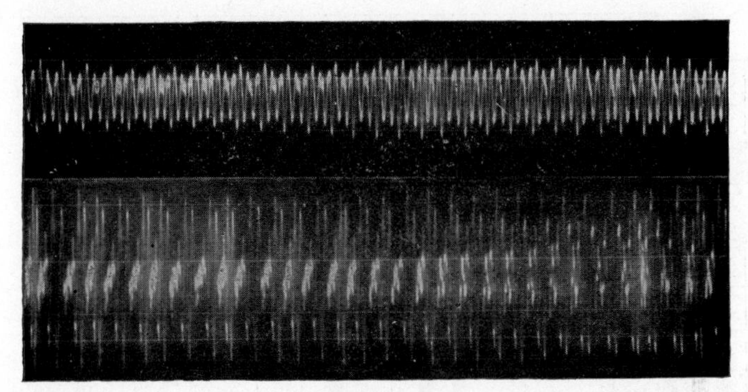

FIG. 13–52. Waveforms of male voice intoning a given vowel in two ways. *Upper record*, "closed" tone; *lower*, "open" tone. The relative amplitude not significant.

the speaking voice. So far as the vowels are concerned the following facts have been definitely established:

1. Any given vowel sound is characterized by the presence of one or more definite groups of partials.
2. The partials which go to make up a given characteristic group are more or less symmetrically disposed about one or more **dominant** partial.
3. The frequencies appearing in these characteristic groups are much the same **regardless of the fundamental frequency.**
4. The various vowel sounds are produced by means of a "modulating" process which consists of so shaping the mouth and the pharyngeal cavities that the proper order and intensity of partials is developed by resonant action.

The sound spectra shown in Figs. 13–53 and 13–54, reproduced from "Speech and Hearing," by Dr. Harvey Fletcher, of the Bell Telephone Laboratories, bear out the first three statements made above. Any particular one of the groups of partials referred to is known as a **formant.** In the spectra shown it will be noted that there are two principal formants, the partials in one constituting a group of relatively low-frequency components, and the other involving a group of relatively high-frequency partials. In general, such a grouping is characteristic of vowel sounds. The details of each

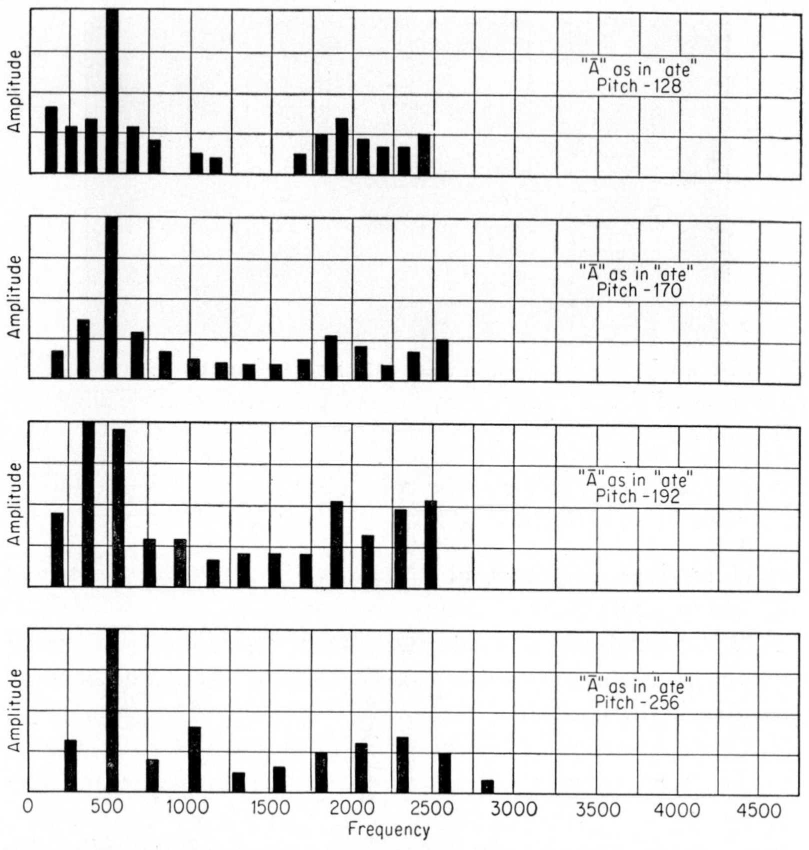

Fig. 13–53. Harmonic content of the vowel $\bar{a}$ when vocalized at three different pitches. (Reprinted by permission from "*Speech and Hearing*" by H. Fletcher, published by D. Van Nostrand Co., Inc.)

formant, and the span of the formants in the scale, determine the voice characteristics of a particular person.

As a result of the extended research Dr. Fletcher found that the mean fundamental frequency of the male voice, when sounding the vowels, is about 124 cps, and for the female 244. The research also disclosed that the **low characteristic frequency** depends upon the particular vowel being uttered, the range being 296 to 955 for the male voices and 332 to 1036 for the female voices in the cases studied. The corresponding values for the **high characteristic frequency** sounds are 1800 to 3000 and 2000 to 3266 respectively.

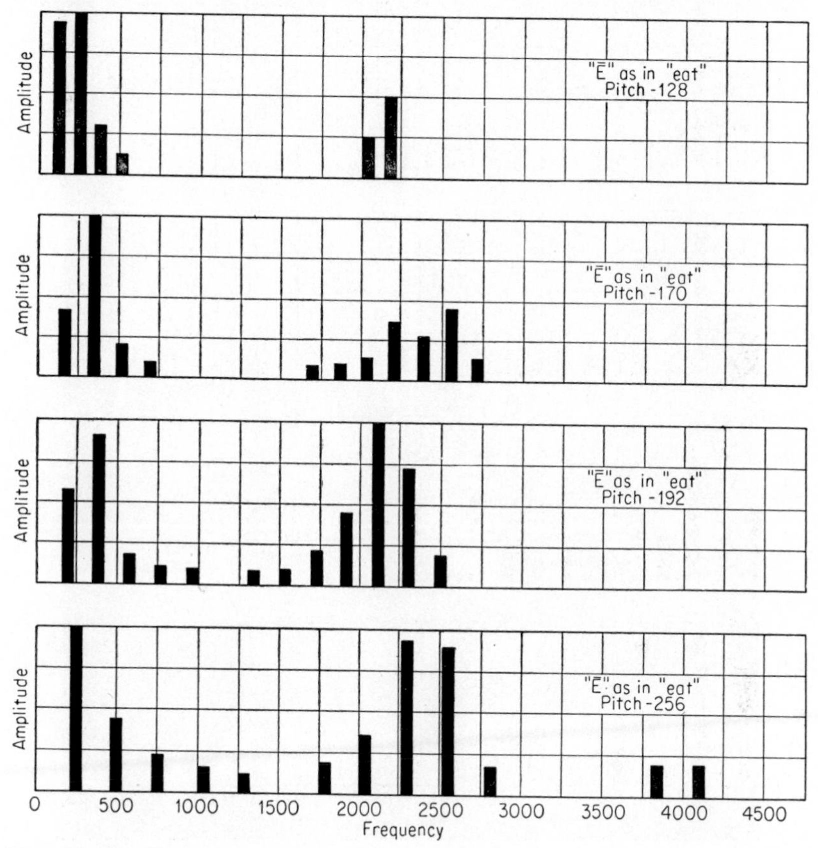

FIG. 13–54. Harmonic content of the vowel ē when vocalized at three different pitches. (Reprinted by permission from "Speech and Hearing" by H. Fletcher, published by D. Van Nostrand Co., Inc.)

The high frequency characteristic components are essential to intelligibility. It is possible to filter out all frequencies below 500 and still secure understandable speech.

In Fig. 13–55 may be seen the waveform of several vowel sounds as intoned by the same person at approximately the same pitch. While the details of a given vowel record will vary with the individual, the general waveform of each vowel is characteristic of

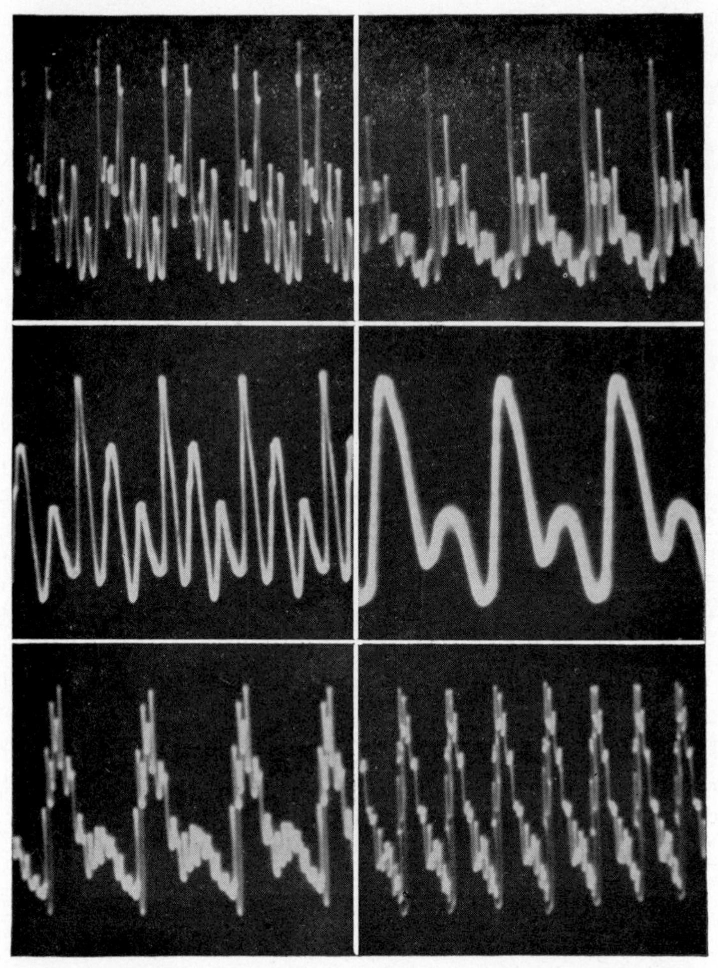

FIG. 13–55. Waveforms of several vowels. *Top row, left, a* as in *ah; right, a* as in *fate; middle row, left, o* as in *oh; right, oo* as in *moon; bottom row, left, e* as in *eh.* Above five records intoned by the same person at approximately the same pitch. *Last record, bottom row, right, e* as sounded by a different person. The spacing and heaviness of the lines not significant.

that particular vocal sound. This statement is based on the examination of many recordings of vowel sounds.

The foregoing observations apply to spoken sounds. But vowels play an even more important part in the singing process. Unfortunately, musical vocalization has not been quantitatively studied to the same extent as speech, but there is reason to believe that the same general relations obtain for vowels which are sung.

In the case of consonants, two types are recognized: in one the sound is formed without vocal cord action, while in the other the cords play a part. The former are called **unvoiced consonants,** typified by *f* and *p*; and the latter **voiced consonants,** of which *v* and *b* are examples. In general, consonantal sounds are impulsive, and hence contain transients. The presence of these transient partials determines, to a great extent, the character of such sounds. The intelligibility of spoken and intoned sounds depends largely on the care with which the consonants are vocalized.

Through proper training it is possible to acquire intelligent and skilled control over the vocal organs and thereby improve the musical quality of the voice. However, much of the so-called vocal training has little scientific basis. For some unaccountable reason, many vocal teachers still refuse to make use of the scientific facilities which have in recent years been made available, and which, if utilized, would greatly facilitate vocal instruction. Technical equipment is now available which, if intelligently and sympathetically employed, would change vocal training from the present cut-and-try procedure to one based on well-known scientific principles. And such a program of training need not in any way detract from the aesthetic and artistic aspects of vocal music. The reader who is interested in voice training will find it both interesting and profitable to read a stimulating and provocative volume by Messrs. Stanley and Maxfield entitled "The Voice: Its Production and Reproduction." In the light of the preceding discussion, Chaps. 2 and 11 of the book just referred to are particularly worth while. Unfortunately many singers are poor musicians. A teacher of vocal music should not only be a thorough musician, but he should also be thoroughly familiar with the fundamental principles of musical acoustics.

A good voice should have a compass of at least two octaves, though many who are classified as singers have a lesser pitch range. The chart shown as Fig. 7–3 indicates the compass of the several types of voice; the limits given are for average voices. As is the case with most musical instruments, the vocal organs usually do not yield the same quality of tone in all registers, but the timbre of any given register can be profoundly modified by training and experience. Figure 13–56 shows the sound spectra of a trained voice singing a note in three different registers. The marked difference in timbre is obvious.

From time to time efforts have been made to produce vocal sounds, particularly the vowels, by artificial means, the object

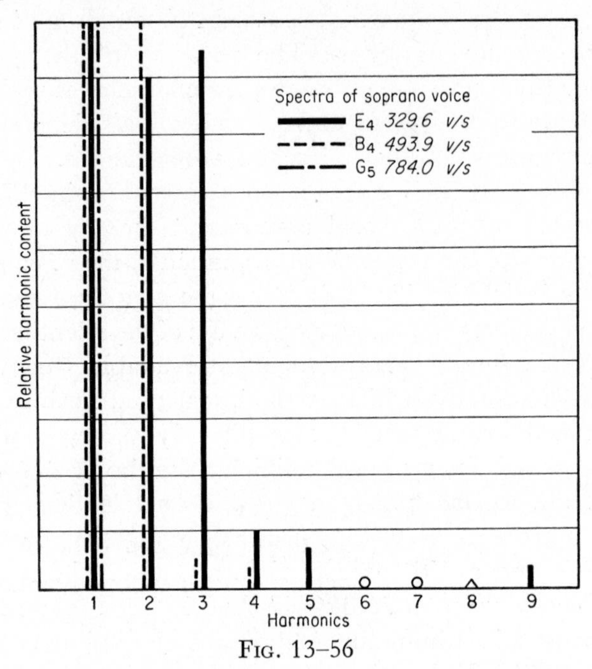

Fig. 13–56

being to acquire further information bearing on the principles underlying the production of sound by the vocal organs. Among the means tried have been tuning forks with associated resonators (Helmholtz); organ pipes (Miller); and more recently, oscillating tubes by the Bell Laboratories. These, and other studies now in progress, will undoubtedly yield useful information which should prove to be valuable in connection with the production of vocal music.

14 *Vibrating Rods and Plates*

14–1. *Longitudinal Vibration of Rods*

In concluding our discussion of sonorous bodies, we shall consider briefly several special cases which are coming to have increasing importance in the field of applied acoustics.

Elastic rods may vibrate in several different ways. The displacements may be longitudinal, transverse, or torsional. Further, the rods may be supported in several different ways. It will therefore be evident that a variety of cases present themselves for consideration. However, only two or three of the possible cases have any present application in the field of music; and we will accordingly confine our attention to those forms of vibration which produce musically useful results.

Before considering these instances, however, it may be worth while to glance at least at one case which at present is chiefly of academic interest only. Reference is made to the vibrations of a rod in which the periodic displacement of the particles is to-and-fro in a direction parallel to the longitudinal axis. Two cases of this type are: (*a*) the rod fixed at one end and free at the other; and (*b*) the rod fastened in the middle and free at both ends. These two cases are sketched in Fig. 14–1. If a metal or glass rod, for instance, is rigidly fastened at one end as shown in (*a*) and stroked longitudinally with a resined cloth, it will be thrown into longitudinal vibration, and will emit a relatively high pitched note. The fundamental note is accompanied by a retinue of extremely high upper partials. The fixed point on the rod is a node and the free end an antinode. Such a rod obeys the same general law as does a closed pipe; the partials occur in the order 1, 3, 5, etc. The frequency is independent of the cross-sectional area so long as this dimension is small compared to the length.

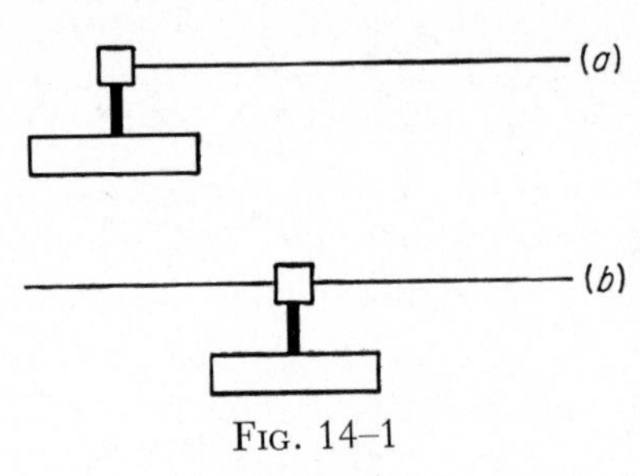

Fig. 14–1

In the case indicated in Fig. 14–1*b*, the rod behaves acoustically in the same manner as an open organ pipe; the fixed point is a node, and each free end is an antinode; hence the wavelength is twice the length of the rod. The vibration of the rod is extremely vigorous; a glass rod will sometimes be fractured, breaking up into annular segments.

One point should be carefully noted in considering the longitudinal vibrations of rods. It is this: the frequency of the vibrations executed by a rod is the same as the frequency of the resulting air wave, but **the wavelength is not.** This will be evident from an examination of a relation which we have used before. In this case

$$f_{\text{rod}} = \frac{s}{\lambda} = f_{\text{air}}$$

Now it is well known that the speed of waves in a solid is not the same as it is in a gas; hence the speed term (s) in the above equation will have one value, say, for metal, and quite another value for air. Therefore λ (wavelength) will be different in the two cases. It therefore follows that

$$\frac{\lambda_{\text{rod}}}{\lambda_{\text{air}}} = \frac{s_{\text{rod}}}{s_{\text{air}}}$$

This is a useful relation; because if we know three terms, we can readily compute the fourth. Advantage is taken of this fact in determining the speed of sound in various solids. In making such a determination, use is made of a piece of apparatus first suggested by Kundt; the setup is shown diagrammatically in Fig. 14–2. A glass tube T held in position by any convenient means contains some light powdered material such as cork dust. A movable piston

P closes one end of the tube. A rod *R* of the material to be studied is rigidly clamped at *C* at its mid-point. By stroking the rod, as suggested above, it may be made to yield its fundamental note. The longitudinal vibration of the rod will communicate its periodic motion to the air within the closed tube. By adjusting the length of this air column by means of the movable piston, a length of tube may be found such that the enclosed air will be in acoustical resonance with the sonorous rod. When this condition obtains, standing waves will be set up in the air, and the dust particles will arrange themselves in well defined heaps along the tube, in the same manner as indicated in Sec. 4–5. By finding the average

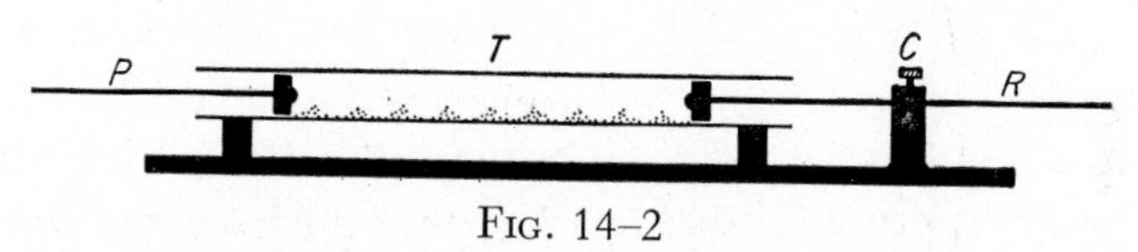

Fig. 14–2

distance between the dust heaps, which correspond to half wavelengths, one can readily compute the wavelength in air corresponding to the wavelength emitted by the rod. Since the speed of sound in air is well known, the speed of sound in the solid sample under test can be easily computed by means of the equation given above. This test is useful in determining the speed of sound in materials which are to be used in the manufacture of musical instruments.

14–2. *Transverse Vibration of Rods*

The transverse vibration of rods follows laws quite different from those which obtain in the cases just cited. Several modes of transverse vibration are obviously possible, but there are only two cases which are of interest from a musical point of view; and the first of these is that in which the rod is fastened at one end and free at the other—sometimes referred to as a fixed-free condition. The case is sketched in Fig. 14–3. Since rods used in musical and related devices commonly have a rectangular cross section, we shall consider only that geometrical form. We have previously dealt with several examples of this type of sonorous body—in the various classes of reeds in connection with certain musical instruments. In those instances the vibration of the reed or reeds was more or less acoustically dominated by the associated resonant air body. In the cases now to be considered, the sonorous body is to be dealt with as an independent vibratile agent, though in certain cases it may be associated with a resonant chamber.

When yielding its fundamental tone, a rod of the type under consideration vibrates as indicated in Fig. 14–3*a*. Rayleigh[1] has shown that the fundamental frequency of a vibrating bar of the type shown is approximately given by the expression

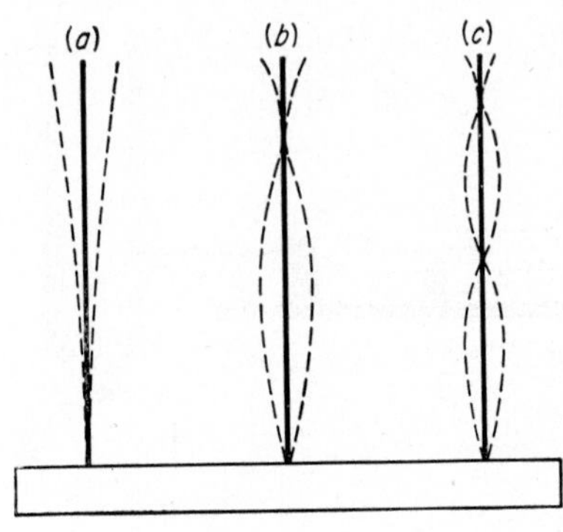

FIG. 14–3

$$f = \frac{kt}{l^2}$$

where t is the thickness parallel to the direction of bending, l the length, and k a constant which in turn involves the coefficient of elasticity of the material of which the rod is made. The thickness at right angles to the direction of vibration is not a factor.

Further, the frequencies of the successive upper partials do not bear the 1, 3, 5, etc., relation to one another as in the case of the stopped organ pipe. Instead, it has been shown by Rayleigh, and confirmed by experiment, that the relative frequencies of the first five upper partials bear the relations given by the numbers 1, 6.25, 17.5, 34.4, 56.5, 84. It is thus apparent that the frequency of the successive overtones increases very rapidly, and that they are inharmonic. Because of their high frequency the partials are quite transient.

As is to be expected from the above-indicated frequency relations, the spacing of the nodes is not equal. In the case shown as Fig. 14–3*b*, the node is located at a point slightly more than a fifth the length of the rod from the free end. In the third case illustrated (*c*) the two nodes are positioned approximately one-eighth and one-half of the length, respectively, from the free end.

The tuning of fixed-free reeds is accomplished by scraping off a

[1] Lord Rayleigh, "Theory of Sound," vol. 1.

portion of the material. The pitch is raised by scraping off a small amount near the free end. To lower the pitch the reed is scraped near the base or fixed end. Examples of the use of fixed-free reeds are to be found in the reed organ, harmonica, clarinet, saxophone, oboe, and bassoon.

The **tuning fork** may be thought of as two fixed-free rods attached to a common base. When vibrating at its fundamental frequency, there are two nodes, one on either side of and close to the point of support, as sketched in Fig. 14–4. That portion of the fork between the two nodes vibrates in synchronism with the two branches; and when attached to a resonant base gives rise to isochronous vibrations therein, thus increasing the intensity of the emitted sound. Immediately after a tuning fork is struck or bowed, one or more of its upper partials can sometimes be heard, particularly if it is struck at a point where one of the antinodes of a partial occurs. Such partials are, fortunately, exceedingly transient; and hence a tuning fork becomes a convenient source of a pure musical tone. The relation given above for the fixed-free bar also gives the frequency of a tuning fork.

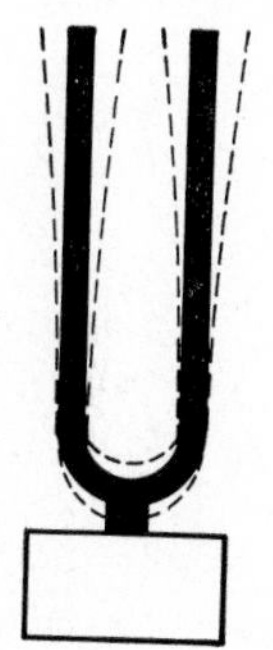

FIG. 14–4

A second important mode of transverse vibration is that in which a rod or bar is free **at both ends,** and in the center. This situation is diagramed in Fig. 14–5. When vibrating at its fundamental frequency, such a rod has two nodes, each node being $0.224L$ from each end, as shown in the sketch. If the bar is supported at these two nodal points and struck in the middle, the fundamental is evoked. The same will be true if it is held in a vertical position by supporting it at one of these points. In cases of this type the fundamental frequency varies inversely as the square of the length. The frequency of the fundamental of a rod free at both ends is not the same as in the case of the same rod fastened at one end, the former being higher and in the ratio of 25:4. In the case now being considered the upper partials are inharmonic, and form a series corresponding approximately to the numbers 1, 2.76, 5.4, etc. It

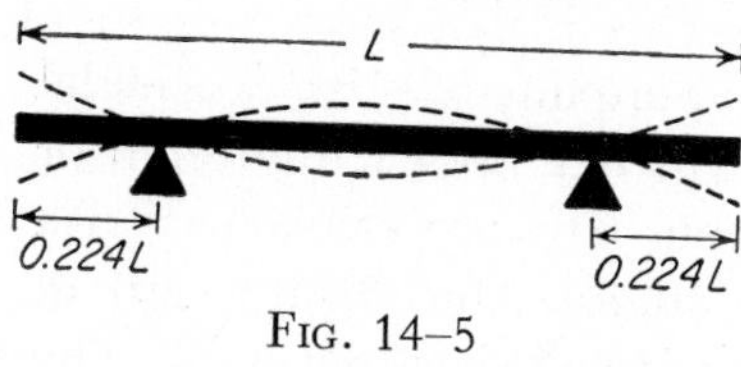

FIG. 14–5

will thus be noted that the partials in this case bear a different relation to one another than in the previous case. Though the overtones do not form an harmonic series, they are not all dissonant, though some are decidedly so.

One use to which the transverse vibration of bars is put is to be found in the percussion instrument we know as the **xylophone.** This organization consists of a series of flat strips of metal or wood. These bars are usually mounted in a horizontal position on soft

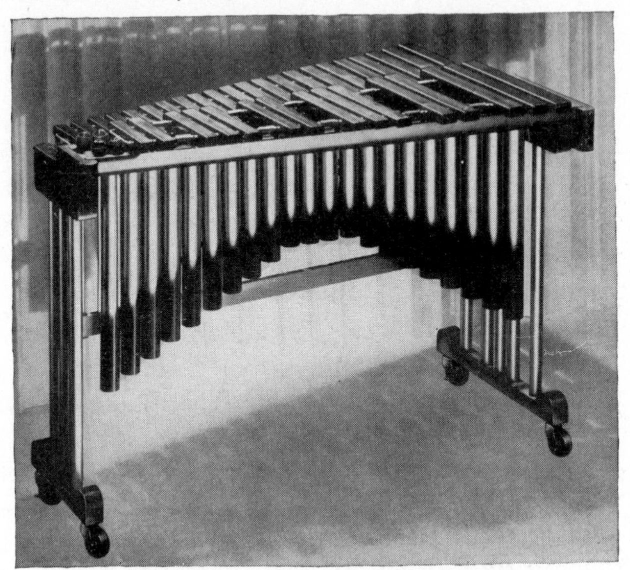

FIG. 14–6. Marimba. (C. G. Conn, Ltd.)

material which touches each bar at approximately its two nodal points. The bars are excited by being struck by a felt-covered or wooden ball. In some cases the sonorous bars are associated with tubular resonators for the purpose of augmenting the intensity of the emitted sound. Such an assembly is called a marimba. The compass is three and one-half octaves, extruding from C_3 to F_7. An illustration of the marimba is shown in Fig. 14–6.

Another use of a generator of the free-bar type is to be found in the chimes incorporated in organs, clocks, and in some door signals, usually consisting of tubular "bars" supported in a vertical position by means of a cord attached at one of the nodal points.

One or two metal bars mounted on resonating chambers are

sometimes used as standards of pitch for use in connection with orchestral tuning.

14–3. *Vibration of Plates*

Chladni, to whose work earlier reference has been made, studied the vibration of plates at great length, and it is to him that we owe the general laws which obtain in such cases.

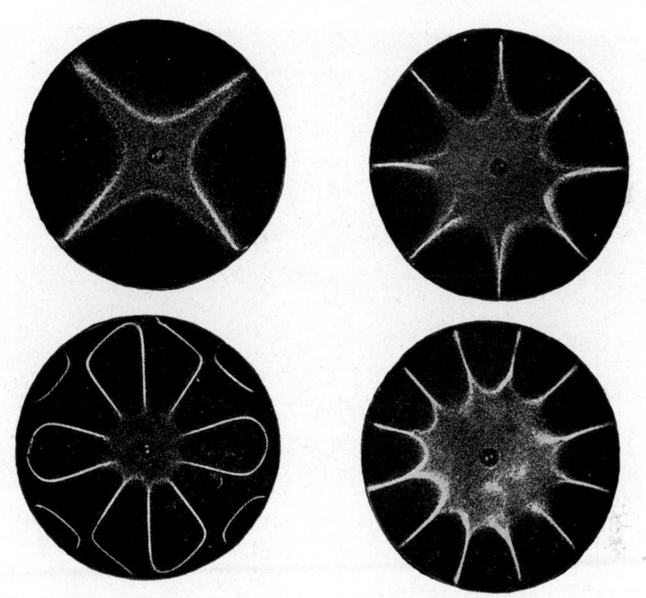

FIG. 14–7. Chladni figures showing several modes of vibration of a round metallic plate. (From "A Textbook of Physics" by C. A. Culver, published by The Macmillan Company.)

If a square or round plate of solid material is clamped at its center and bowed at some point on its edge, it will emit a musical note which is characterized by a retinue of upper partials which are largely inharmonic. Such a plate breaks up into a series of acoustical segments, as in the case of strings and bars. The simplest mode of vibration, that is, when the plate is giving its fundamental note, is that in which there are four segments and four corresponding nodes, as shown in the upper left diagram of Fig. 14–7. The series of Chladni figures, as they are called, shown in the illustration, were made by photographing the pattern made when the surface of a vibrating metal plate is partially covered with fine sand; the sand arranges itself along the radial nodal lines. By damping the edge at some point and bowing at another, various segmental patterns, and corresponding tones, may be attained, as shown in the figure. It is important to note that adjacent segments are always moving in opposite directions at any given instant. An oscillograph of the tone emitted by the plate used in making the above Chladni

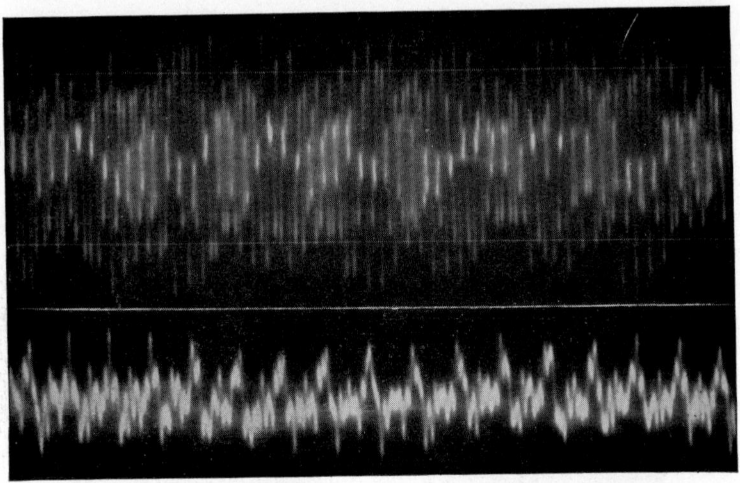

FIG. 14–8. The upper trace shows the waveform of a vibrating plate recorded immediately after being excited. The lower record was made after the plate has been vibrating for a few seconds. Note the reduction in the number of partials.

patterns is reproduced in Fig. 14–8. A record of the sound emitted by the same plate, taken about a second later, is shown in the lower of the two waveforms. Note the marked decrease in the intensity of the higher partials.

Chladni's laws of vibrating plates are: (1) The frequencies of two plates of like shape, and showing the same nodal patterns, vary as the thickness of the plates. (2) The frequencies of two plates of the same thickness, and showing the same nodal patterns, vary inversely as the square of their diameters. The two foregoing statements may be combined into a single law having the form: **If the plates are similar solids, the frequencies will vary inversely as**

the homologous dimensions. For instance, if one plate, either square or round, is twice as thick and has twice the diameter of another, the larger plate will yield a note which is an octave below that given by the smaller body. As a matter of fact, this law holds good not only for solids but also for liquids and gases. It is, in fact, a general law of sonorous bodies.

While vibrating metallic plates are not used directly for the production of music, such vibratile bodies are widely employed in connection with the reproduction and the recording of musical sounds. In such devices the plates usually take the form of a thin sheet or diaphragm. Musical sounds are picked up by some form of microphone and by one of several means converted into corresponding alternating currents. These in turn are amplified and fed into a loudspeaking system, or are photographically recorded on a cinema film. In most cases the so-called microphone consists of a diaphragm which is mechanically associated with an electromagnet system or a pressure-sensitive crystal (piezoelectric) in such a manner that the variable air pressure on the diaphragm caused by the sound wave gives rise to corresponding minute electric currents. Each such diaphragm has one or more natural periods of vibrations. If the inherent fundamental period of a microphone diaphragm should chance to coincide with one of the frequencies of a complex incident sound wave, that particular frequency, because of the resonance effect, would be unduly augmented by the electro-acoustical system constituting the sound pickup device. The result would be that the reproduction of the incident sound would be defective; in other words, acoustical distortion would obtain. For instance, the diaphragm of an ordinary telephone transmitter has a natural period in the region of 800 cps. In order to avoid this, one of two means is usually adopted as a remedial measure: Either some method is used to damp the diaphragm somewhat at its natural frequency; or the diaphragm is stretched so that its natural period is very high—above that of any frequency which is likely to exist in the incident sound.[1]

[1] Sound is sometimes received on a piezoelectric crystal without the use of a diaphragm and thereby converted directly into an electric current. See the author's text on "Theory and Applications of Electricity and Magnetism," p. 232, for a discussion of the piezoelectric effect.

14–4. *Bells*

Plates are acoustically related to bells, and bear somewhat the same relation to them as rods do to tuning forks. A bell may be thought of as being derived from a disk, curved in shape and loaded in the middle. A cross-sectioned view of a musical bell is shown in Fig. 14–9. The segmental vibration of bells, in general, is the same as in the case of plates; the number of vibrating segments is always even, and the fundamental, or ground tone, is emitted when the bell breaks up acoustically into four segments. The nodes may form both radial and circular lines, the first running up and down the bell and the latter around it. The first-mentioned lines are referred to as nodal meridians and the second as nodal circles. As in the case of plates, adjacent segments move in opposite directions at any given instant. Bells are made of a special alloy known as bell metal, containing three to four parts of copper to one part of tin. The frequencies of bells vary inversely as their homologous dimensions. It follows, therefore, that the pitch of bells of the same shape and material varies inversely as their diameters. It is also true that the pitch of the notes emitted by bells varies inversely as the cube root of their weights—the larger and heavier the bell, the lower the pitch.

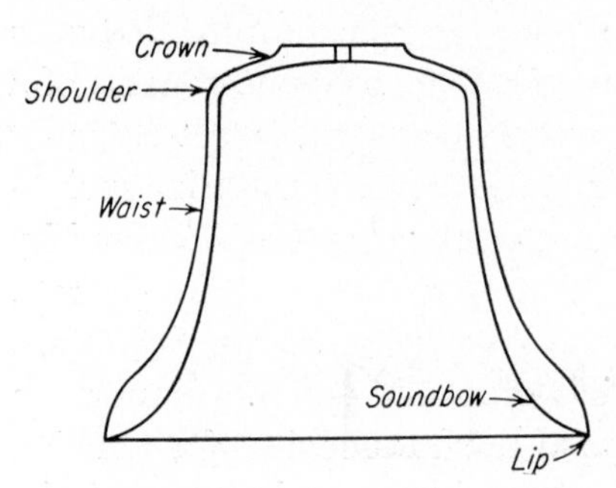

FIG. 14–9. Cross section of a bell.

The art of bellmaking is very old; van der Gheyn (1550) and Hemony (1650) were successful in casting bells which have since served as models for later artisans in this field. These craftsmen did for bellmaking what Amati and Stradivari did for violins. As a result of his study and experience, Hemony concluded that a good bell should have three octave notes, one major, and one minor third. It is now felt that, in order to develop a pleasing tone, a bell should have the partials indicated in the diagram shown in Fig. 14–10. The minor third is considered to be a particularly important component, giving to the bell tone its plaintiveness, and accounting for its appeal quality.

The pitch of a bell, as heard by the listener, is not determined by the lowest or hum tone, but by a component that is higher in pitch than the hum tone. This partial appears to be an octave below the fifth, and, strangely enough, has no objective existence; it appears to be an aural harmonic. Notwithstanding the fact that this subjective partial decreases in intensity more rapidly than the hum note, it is the musical component by which the pitch of a bell is commonly designated. Actually there are, in many cases, a number of high partials in a bell tone that are inharmonic. Fortunately, however, most of these are quite transient and hence do not seriously detract from the quality of the tone. In Fig. 14–11 is to be seen the waveform of a particular relatively small bell. The record begins immediately after the bell was struck. As in the case of the plate the sound contains many upper partials. About two-tenths of a second after the bell is sounded a characteristic

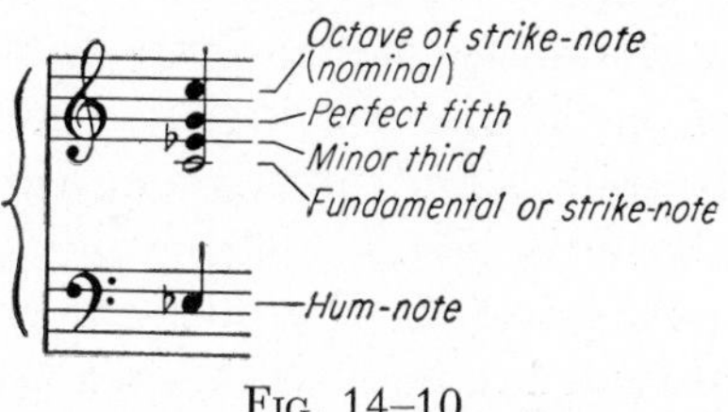

Fig. 14–10

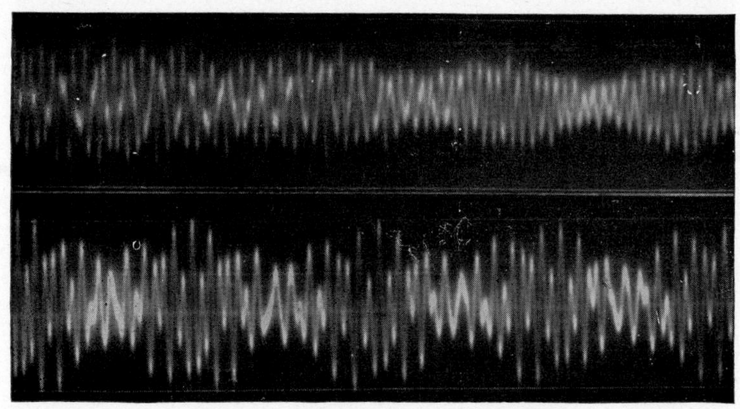

Fig. 14–11. Waveform of a bell.

pulsating tone is often heard. This vibration in intensity is due to the existence of beats, as indicated at the right of the record. The lower tracing is an enlargement of that portion of the record. Such beats are probably due to a slight difference in pitch between the vibrating segments, due in turn to nonuniformity in mass distribution. The late Professor Jones, of Smith College, who made an extensive study of bell tones, found that immediately after being struck the fifth partial was the strongest component but that after two or three seconds the minor third was the dominant overtone.

When bells are cast at the foundry they are made slightly larger than they will be when finished. When being tuned the cast bell is placed on its crown on a revolving table. By means of a cutting tool shavings are cut from the inside until the desired complement of partials has been secured. Removing metal from the rim raises the pitch. The pitch is lowered by removing some metal at places above the rim.

A group of bells designed to yield chromatic intervals and arranged to be struck from a keyboard is known as a **carillon.** The number of bells involved in such an assembly varies from 23 to 72. Such musical bells are hung "dead," and the larger installations have a total weight of many tons. The bells of a carillon are tuned to the equally tempered scale, but because of the inharmonic relations of some of the partials only a limited number of chords are feasible. Usually only the melody is played. With the larger installations, however, a melody can be played on the larger bells while an accompaniment can be executed on the smaller units. Much musical knowledge and mechanical skill are required to produce satisfying results. One who plays such a set of musical bells is known as a **carillonneur.**

It is now possible to duplicate the sound of carillon bells by electroacoustical means, and thus avoid the necessity for the strong towers needed to support the traditional carillon installation. The cost of an installation of this type is also substantially less than for the cast-bell type of assembly. This newer method of producing bell-like tones will be described in the following chapter.

14–5. *Drums*

Another musical application which involves certain of the principles connected with vibrating plates is that of drums. Here we are dealing with membranes, usually circular in shape, and often associated with a resonant air body.

The use of drums dates back to the dawn of history—probably

before, and the use of this type of percussion instrument is common to all peoples. Modern drums fall into two groups: (1) those designed to produce sounds of definite pitch and timbre such as the kettledrum used in the orchestra; and (2) those which have no definite pitch, and which are used for marking rhythm or for the production of certain special effects.

The **kettledrums** or **timpani** (Fig. 14–12) may properly be classified as musical instruments. In construction they consist of a hollow hemispherical body of metal over the open end of which is stretched an animal membrane, this vibratile member being under tension. By means of a number of screws arranged around the edge, the tension can be adjusted and the pitch thereby controlled for a range of approximately a fifth. The kettledrum emits a musical note of definite pitch. An oscillographic record of this sound is given in Fig. 14–13. A stretched membrane vibrating alone gives rise to certain upper partials which succeed each other at very close intervals—in some cases less than a semitone. When, however, such a membrane is associated with a suitable resonant cavity, such as is the case in the drum just mentioned, the tone color is greatly

FIG. 14–12. Kettledrum.

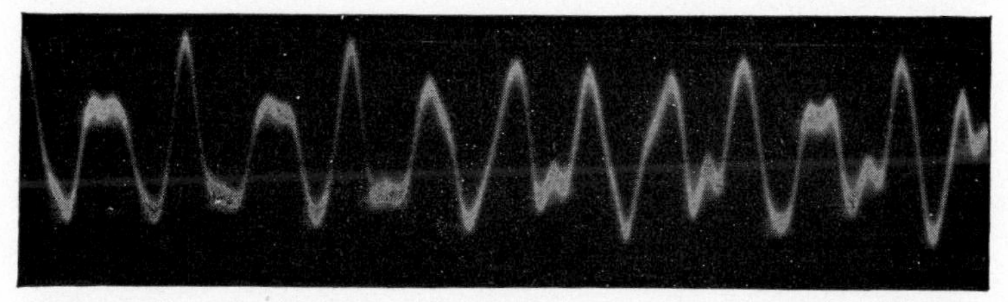

FIG. 14–13. Waveform of kettledrum. Note change of phase.

modified, and we have a timbre which has the characteristics shown in the illustration. Three kettledrums are now commonly employed in orchestral work. One is tuned to the tonic of the key; another to the subdominant; and the third to the dominant. The smallest drum has a range extending from $B_2\flat$ to F_3; the compass of the largest extends from F_2 to C_3.

Solos for timpani appear frequently in symphonic compositions, as in Rimsky-Korsakov's *Russian Easter*.

As indicated above, the bass and snare drums do not emit a definite pitch; neither do they present a definite tone quality. They will therefore not be considered further.

15 *Electronic Musical Instruments*

15–1. *Electronic Carillon Systems*

It is possible to produce carillon effects by electroacoustical means. One method of attaining this end makes use of the chimes that commonly form part of church organs. The sound from these chimes can be picked up by a suitable microphone, amplified, and radiated from a group of stentors (loudspeakers) installed in the tower of the church. The sound from such a tower-chime system partakes of the nature of bell tones, and is found to be quite pleasing. However, in order to secure satisfactory results from such a tower-chime installation it is important that the following general specifications be followed:

1. The organ chimes should be of the highest quality.
2. The chime assembly should be housed in a room that is thermostatically and acoustically controlled.
3. The microphone should be a high-grade unit of the electrodynamic type, and properly positioned with respect to the chime assembly.
4. The amplifier should be free from distortion effects when operated at its rated capacity of 100 watts.
5. The four stentors should be of the reentrant horn type, and each should be capable of handling 25 watts without distortion.
6. The horns should be mounted at a suitable height.

Another substitute for the conventional carillon bell assembly makes use of a series of slender bars fastened at one end, as sketched in Fig. 15–1. A series of such free-fixed rods of suitable lengths is rigidly fastened to a heavy base and so arranged that they may be excited by means of an electromagnetic striking mechanism which in turn is controlled by the keys of a special or a standard console. Such rods yield a fundamental and several upper partials. One or more electrostatic or magnet pickup devices are supported very near each tone rod. When a given rod vibrates, a minute electrical

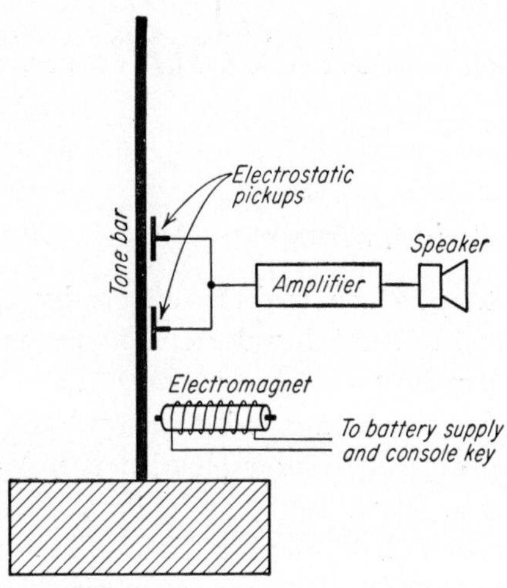

FIG. 15–1. Sketch showing essential components of one form of electronic-bell assembly.

current is developed in the pickup unit and this small current is amplified electrically and in turn fed into suitable loudspeakers. By careful design of the tone bars and the position of the pickups, those harmonics which are characteristic of cast bells may thus be made available.

Electrocarillon bell assemblies, such as the one last described, are commercially available. They give musical results that are comparable with those secured from the conventional carillon installations. The cost of either of the types above mentioned is only a fraction of the expense that is involved in providing a cast-bell assembly.

15–2. *Electronic Organs*

During the past few years several attempts have been made to develop an instrument that would simulate the tones of a conventional pipe organ, and several designers have attained a considerable measure of success.

Two different lines of attack have been followed in attempting to solve this problem by electronic means. The first method involves the direct synthesis (Sec. 8–6) of a given tone color. The synthesis of any sound may be accomplished by generating sinusoidal electric currents and then mixing such elemental components in the proper proportions to produce the desired complex waveform.

A second method of approach is based on the production of an electrical current containing a multitude of frequencies and then, by means of an electrical filter system, attenuating or emphasizing various components in such a way that the resultant waveform corresponds to that of the desired sound. It was pointed out in Sec. 8–4 that any sound may be analyzed into its components. Organ tones have been so analyzed. Knowing the sound spectrum of a given organ tone it is possible, by either of the above methods, to duplicate the tone of such a pipe. Theoretically the case is quite simple, but to combine the various synthesized tonalities that enter into an organ chord involves rather complicated networks.

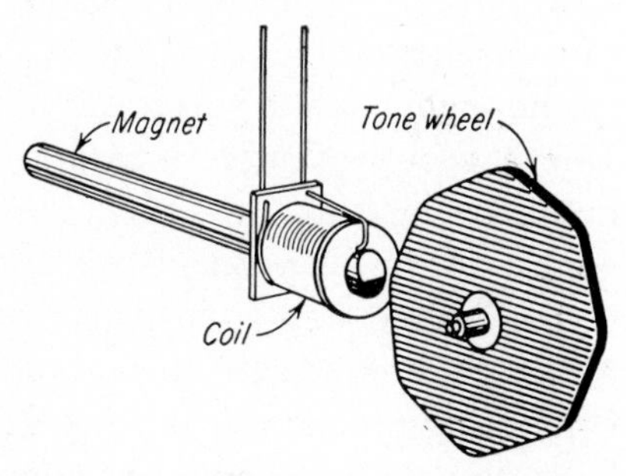

FIG. 15–2. (Hammond Instrument Co.)

Referring to the first method mentioned it may be said that there are several means whereby simple sinusoidal currents may be generated. One procedure makes use of what, in electrical parlance, is known as an inductor type of a-c generator. The essential parts of a generator of this type are depicted in Fig. 15–2. The unit consists of a small permanent magnet about one end of which are wound a few turns of wire, as shown. In front of one end of this magnet there is caused to revolve at constant angular speed a small disk of soft iron about 1 in. in diameter. The periphery of this *inductor* disk has a wavelike contour of definite form. As the disk revolves, being of magnetic material, the irregular edge causes the strength of the permanent magnet to change in value; and this change in magnetism will conform to the waveform described by the periphery of the rotating disk. By proper design a periodic change in magnetism closely resembling a sine wave can be secured. This cyclic change in the strength of the magnetism will induce in the associated winding a minute electric current. This feeble electric alternating current, then, becomes one of the ninety-odd similar sources of energy from which any given musical tone is synthesized. The angular speed of the inductor disk and the number of sinusoidal projections on its edge, together, determine the frequency which that particular unit will generate. By having available a group of such generators, and by the use of appropriate auxiliary circuits, it is possible to mix the output from several generators in the proper proportion to produce an electrical replica of the desired tone. This weak complex electric current is fed into an amplifier which in turn actuates some form of transducer, thus converting the electrical output into sound.

Some comprehension of the complexity of the electrical circuits involved in the production of synthetic music may be gathered from an examination of the diagrammatic sketch shown in Fig. 15–3. The diagram is intended to represent a very small part of the possible electrical circuits (very much simplified) of an "electronic" organ. The circles at the left (with curved lines enclosed) represent a series of electrical generators, each such generator being arranged to develop a minute electrical current of definite and fixed frequency, as indicated by the adjacent note letters. Several tonics and one full octave are sketched. Commonly there are 96 such generators, thus providing for a fundamental and the first seven upper partials. It is the judgment of the designers of such instruments that the absence of overtones above the seventh will not seriously impair the quality of the desired tone.

Suppose it is desired to produce a diapason tone whose fundamental is C_1, as shown. By means of the diapason stop connections are made to the generators indicated, and appropriate amounts of the 2d, 3d, and 4th harmonics (in the form of minute electrical currents) are combined into a complex electrical waveform characteristic of the diapason tone. When the player depresses the key K_1, this complex electrical current is passed to a suitable amplifier and thence to a loudspeaker, where the electrical current is converted into sound.

Suppose that the organist desires, say, to play the octave (C_3) on the flute stop at the same time that he is sounding the diapason C_1. In order to do this he presses key K_2. The flute stop is connected to generator C_3 as a fundamental and also, for instance, to C_4 as

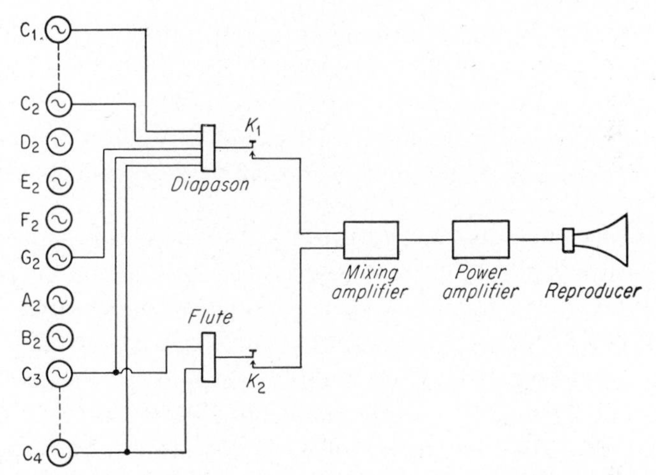

FIG. 15–3. Schematic diagram of one form of circuit utilized for the purpose of synthesizing musical sounds.

the first upper partial (second harmonic). It will be noted that generator C_3 may serve as the fundamental of the flute stop and also as the third overtone of the diapason. This process of causing any given generator to function as a fundamental of one stop and as an upper partial of another is known as "borrowing." The closing of K_2 simultaneously with K_1 will cause electric current from the diapason and the flute stop to be fed to the mixing amplifier at the same time, with the result that one will hear both complex tones from the loudspeaker simultaneously, as would be the case on the conventional organ. There is a basic difference, however; in the standard organ the sound which we hear is generated as sound by the pipes; in the synthesizing instrument electrical impulses are generated which in turn are converted into sound by means of thermionic tubes and the associated electromechanical organization. Because electronic tubes are sometimes used in generating the minute electrical current needed, and also in the conversion of such periodic electric currents into sound, such assemblies are referred to as electronic instruments. A complete organ would involve from 20 to 35 stops and possibly two or more manuals. With this in mind, it will be apparent that the electrical organization necessary in order to produce music by the process of synthesis is of a highly complex nature.

In a second method of producing organ tones synthetically, use is made of an electrostatic generator. In Sec. 8–6 a brief reference was made to the electrostatic generator for the design of which the author is more or less responsible. Dr. Floyd A. Firestone has also designed an electrostatic generator that may be utilized for tone synthesis.

An electrical condenser consists of at least two parallel conducting plates, usually metal, separated by a very small intervening space of nonconducting material, such as air, mica, or waxed paper. We say that a pair of such conductors has *capacitance*—the ability to store electrons. If and when the plates of a condenser are first connected to a source of electrical potential difference, such as the terminals of a dry cell, there will be a rush of electrons from the battery to the plates of the condenser—the plates become electrically *charged*. This movement of electrons constitutes a momentary pulse of electric current. When a given potential difference is impressed on the plates the magnitude of the charge acquired by the condenser will depend upon the amount of the overlapping plate area, that is, upon the capacitance of the condenser system.

If one of the plates of the condenser has a particular shape, and the other plate is caused to move parallel to the first member, the capacitance will vary in value as the overlapping area changes. If, for instance, we have a metal plate P_1 having the shape shown in Fig. 15–4, and if we cause another plate, shaped as P_2, to rotate very near to and parallel with P_1, the capacitance of the system will vary in a manner determined by the shape of the rotating plate. Under such circumstances the charging current will vary in a corresponding manner. If the rotating plate (the rotor) is properly shaped, the capacitance will vary sinusoidally as the rotating member moves, and the charging current will have the same waveform. The frequency of the current variations will depend upon the angular speed of the rotor and also upon the number of sectors that make up the rotor and the stator.

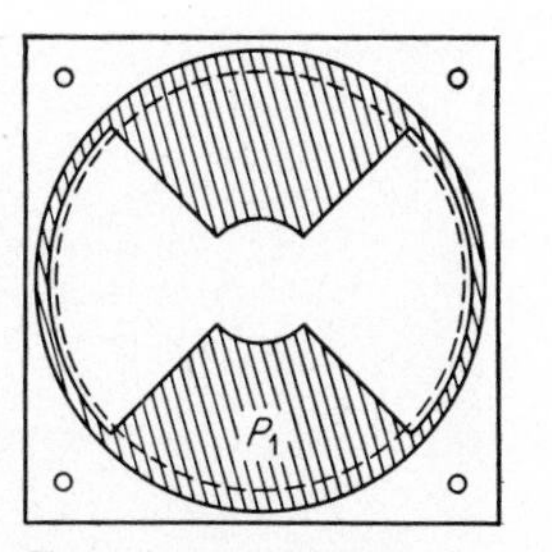

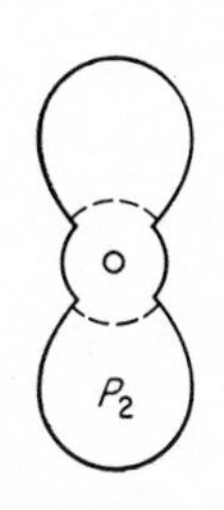

FIG. 15–4. Stator and rotor electrodes of electrostatic generator.

If the varying current is caused to pass through a resistor (R in Fig. 15–5) the potential difference developed at the terminals of R becomes a source of pulsating electrical energy. The small amount of energy thus made available may be amplified by means of a suitable electrical network, as indicated in the diagram shown in Fig. 15–5. Figure 15–6 shows the actual waveform of such an electrostatic generator. Thus we have an elemental generator that will perform the same function as the inductor type of generator previously described. In Fig. 15–5, G_1 and G_2 represent

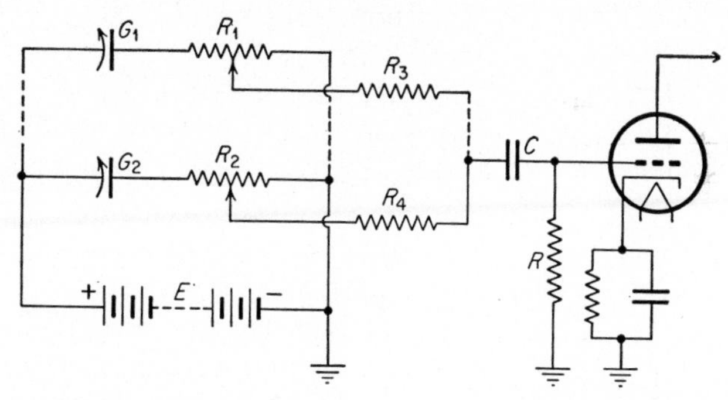

FIG. 15–5. Circuit diagram showing a group of electrostatic generators and associated amplifier connection.

a number of such electrostatic generators, the outputs of which may be fed into the amplifier, and thus build up a given tone color.

It may be added, in passing, that the sectors of the stator of an electrostatic generator may be so shaped that waveforms other than sinusoidal may be developed. The leading British electronic organ makes use of electrostatic generators having specially shaped stator plates. The largest model of the British instrument is a three-manual assembly incorporating 93 stops.

Elemental sinusoidal waveforms may also be generated by means of electronic tubes. A tube, such as those found in any radio

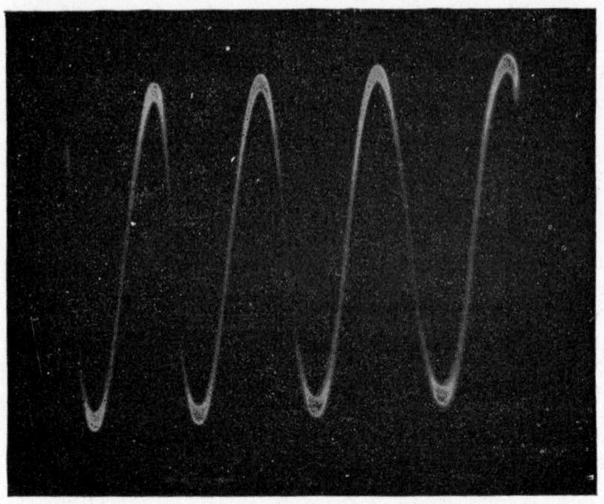

FIG. 15–6. Basic waveform produced by the synthephone.

receiving set, will, if connected into a suitable electrical network, function as a generator of sinusoidal currents. Such a circuit is sketched in Fig. 15–7. (For a detailed discussion of tube oscillators see the author's text entitled "Theory and Applications of Electricity and Magnetism," p. 524.) A group of such generators can be incorporated in the organization shown in Fig. 15–3, and an electronic organ developed on that basis. Such an assembly has the advantage that there are no mechanically moving parts.

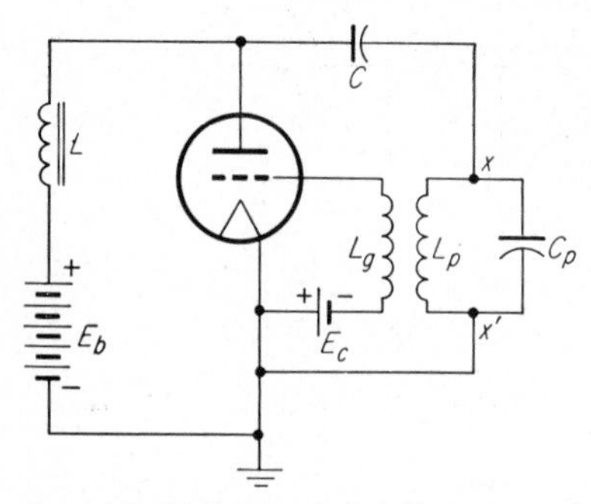

FIG. 15–7. Schematic diagram of tube-generator network.

From the foregoing outline it is evident that there are at least three methods whereby sinusoidal currents may be generated for use in connection with the synthesis of musical sounds.

We next proceed to examine what might be considered to be a selective method of developing the desired waveform. In our study of the generation of violin tones (Sec. 11–1) it was noted that a saw-toothed type of oscillation is always made up of a large number of sinusoidal components. With a generator of this general type available it would be possible, through the aid of filters, to select a group of partials that correspond to a particular musical tone. There are at least two methods that may be followed in developing a saw-toothed wave for use in the domain of electroacoustics. One method utilizes the electrical oscillations that may be produced by an electron tube. Above it was pointed out that such a tube could be made to develop a sinusoidal wave output. It is also true that a tube of this character may be caused to yield an output whose waveform is of the saw-toothed shape. Then by utilizing a filtering process the desired waveform may be made available for use in an organ of the electronic type. Several of the electronic organs now on the market are constructed on this principle.

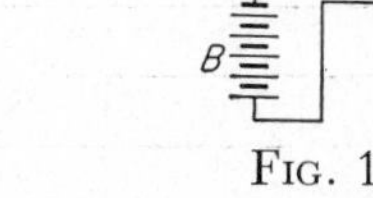

FIG. 15–8

One well-known electronic organ makes use of a vibrating reed to secure the desired composite waveform. It will be recalled that the reed of a clarinet, for instance, develops, when sounded alone, a large group of partials. One manufacturer makes use of wind-blown metallic reeds, similar to those used in the American reed organ, as one member of a variable condenser. Now the capacitance of a condenser depends not only on the size of the plates, as previously indicated, but also on the distance between the two plates, and it is this factor that is made variable in the case of the reed generator. A diagrammatic sketch showing the essentials of such a reed generator is to be seen in Fig. 15–8. The vibrating reed V forms one plate of the condenser, and the small metal plate P the other component. These two condenser electrodes are charged by means of a battery of 100 to 300 volts. As the electrostatic capacitance of the condenser, formed by V and P, changes, owing to the vibration of the reed, the varying charging current will flow through the resistor R; which in turn will establish a varying potential difference between the terminals of that resistor. Thus the resistor becomes a source of alternating electrical energy, as in the case of the electrostatic condenser previously described.

By proper design of the reed and the other associated components, this reed type of electrostatic generator will develop an electrical output that is rich in harmonics. The sound of the reed is not utilized in any way; indeed, it is suppressed. Again by utilizing the principle of electrical filtering the desired form is secured, and in turn transformed into sound of the desired timbre, by the methods previously described. The instrument makes use of a series of such reed generators each of which develops a particular waveform when associated with a suitable electrical filter. It is to be noted that this

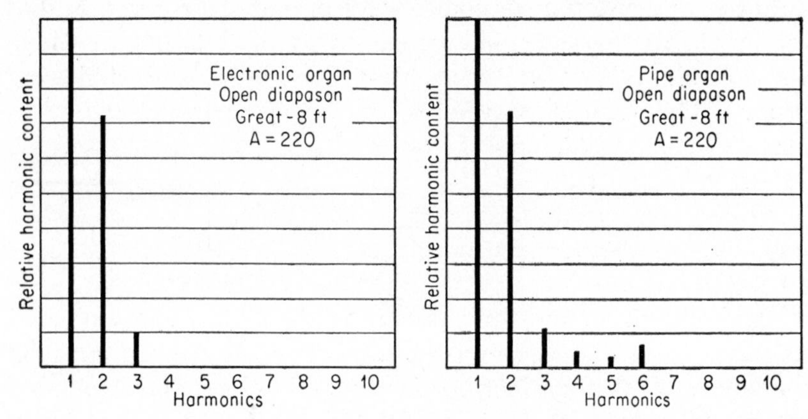

FIG. 15–9. Comparison of a certain note when sounded on a particular electronic organ with the same note when played on a conventional organ.

instrument, and the one employing a multiple-frequency tube generator, does not utilize the principle of synthesis.

Marked advances are being made in the design and construction of electronic organs, but it is probable that this type of instrument will never completely displace the traditional pipe organ. However, the new type of organ is coming into wide use, owing in part at least to its comparatively low cost. The spectrum of one tone on one of the well-known electronic organs is shown in Fig. 15–9. In the same illustration is also to be seen the corresponding tone spectrum of one of the better pipe organs. Note the difference in the harmonic content. It is of course possible to make the electronic instrument have exactly the same harmonic content as the conventional organ, if so desired.

For a detailed technical discussion of electronic organs the reader is referred to a recent book entitled "Electronic Musical Instruments" by Richard H. Dorf, and published by Radio Magazines, Inc., Mineola, N.Y.

15–3. *Electronic Piano*

Various attempts have been made to develop a piano in which the sounding board, which functions as a mechanical amplifier, would be replaced by an electrical amplifier. Theoretically such a substitution is possible, but in practice a number of difficulties present themselves.

It is, for instance, theoretically possible to synthesize a piano tone as is done in the electronic organs first described above. Such a

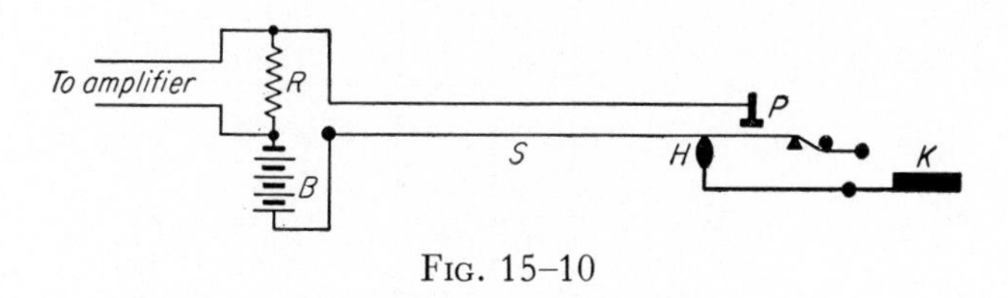

FIG. 15–10

hypothetical instrument would not involve the usual strings or sounding board. However, such an approach to the problem would, in the final analysis, necessitate an assembly that would be as large as, if not larger than, the ordinary piano, and the cost would probably be comparable.

It has been suggested that it would be possible to make use of an electromechanical setup somewhat like the arrangement shown in Fig. 15–10. According to such a plan *S* represents a string (or group of strings) corresponding to some particular note on the piano. The usual key and felt-covered hammer are indicated by *K* and *H*, respectively. Placed near to the string, at an appropriate point in its length, would be a small metal plate *P*. The steel string and the plate would act as the two elements of an electrical condenser, and the rest of the story would be the same as that outlined in the case of the electronic carillon (Sec. 15–1). Of course an electromagnetic pickup could be substituted for the electrostatic unit. But there is one serious difficulty connected with such a suggested layout. The vibrations of a piano string are complex, consisting of the fundamental and a varying group of upper partials. Tests made by the author show that the waveform "picked up" at various points along such a string differs appreciably in character. Evidently, several pickups would be required for each string if one were to capture the true tone picture of the string, and, further, the phase relations between the output of the several pickups would present an additional problem. Obviously the problem of developing an electronic piano is a complex one, but the difficulties are not insuperable. In time such an instrument may be developed.

16 *Recording and Reproduction of Music*

16–1. *Phonographic Recording*

In view of the fact that musical recordings are widely used, it will be in order to examine, briefly, the more common methods of making recordings, and the processes by which such records are reconverted into sound.

The oldest type of recording is that which we know as the phonograph record. As originally developed by Edison, the record consisted of indentations made in a sheet of tin foil wrapped around a

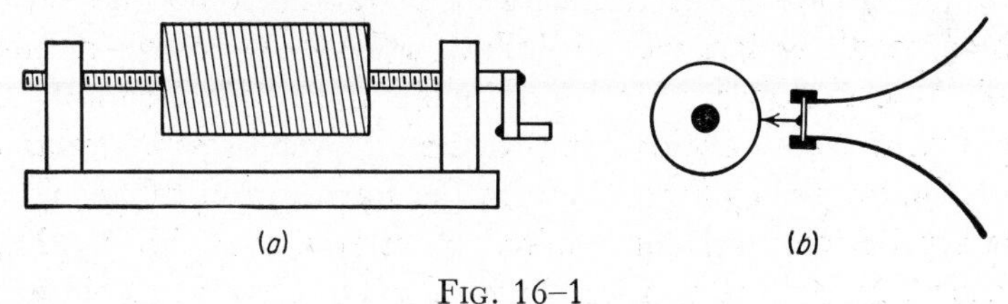

Fig. 16–1

cylinder. The original recording mechanism was simple. It consisted of a grooved cylinder mounted on an axis the protruding ends of which were threaded, as sketched in Fig. 16–1*a*. It will be evident that by mounting the cylinder as shown, it could be rotated and moved forward simultaneously. A thin diaphragm carrying a stylus, mounted at one end of a horn (Fig. 16–1*b*), was supported in such a position that its free point touched the surface of the metal foil above the groove on the surface of the cylinder. The varying pressure of the sound waves entering the horn caused the diaphragm to vibrate, and in so doing produced radial indentations in the foil. If, then, the motion were retraced, the needle, following the groove originally made, would cause the diaphragm to vibrate, thereby reproducing the original sound. Later a wax surface was substituted for the metal foil, with improved results.

The next improvement in sound recording consisted in the introduction of a flat disk made of moderately soft material in which a sound track was formed by means of a laterally displaced cutting stylus, the disk being rotated by means of a spring-driven motor. The vibratile member D (Fig. 16–2) usually consisted of a thin disk of mica. Mechanically articulating with this was a lever, L, the lower end of which was shaped in the form of a cutting edge. As sound waves actuated the diaphragm the recording end of the lever cut a wavy groove in the soft material composing the disk. An electrotype copy of this original recording was then made, and this became the "master" record; from it large numbers of replicas were stamped.

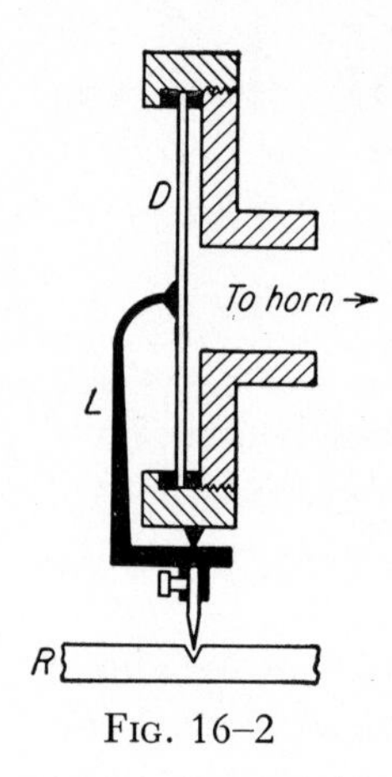

FIG. 16–2

The "playback" was accomplished by replacing the cutting tool by a needle which, when it followed the wavy groove, actuated the vibratile member, thus reproducing the original sound. This arrangement was a decided improvement over previous methods, though the frequency limits were rather narrow and the mechanics of the reproducing system resulted in relatively rapid wearing of the record. Nevertheless, recordings of some of the great singers of the time were made by means of this type of equipment.

The wholly mechanical method described above has, in recent years, been succeeded by an electrical method. This improvement consists in actuating the cutting stylus by means of a small electromagnet, as shown in Fig. 16–3*a*. A microphone, M, serves to convert the energy of the sound waves into a weak variable electrical current. This current is an electrical replica of the original sound waves. The small current thus generated is amplified by means of electronic tubes. The amplified current thus reproduced, when passed through the winding of the electromagnet *W*, results in a magnetic field which varies in intensity in conformity with the output of the amplifier, *A*. The varying magnetic field causes the soft iron disk, *D*, to be attracted, in varying degree, to the pole of the electromagnet. This, in turn, actuates the pivoted cutting stylus, *L*, as indicated in the sketch.

The method of reproduction consists in reversing the foregoing process, as diagramed in Fig. 16–3*b*. A playback needle follows the groove, thereby causing the iron disk, *D*, to move harmonically toward and away from the pole of the electromagnet. This in turn results in the development of a small variable electrical current in the winding. (Such a reproducing assembly is known as a

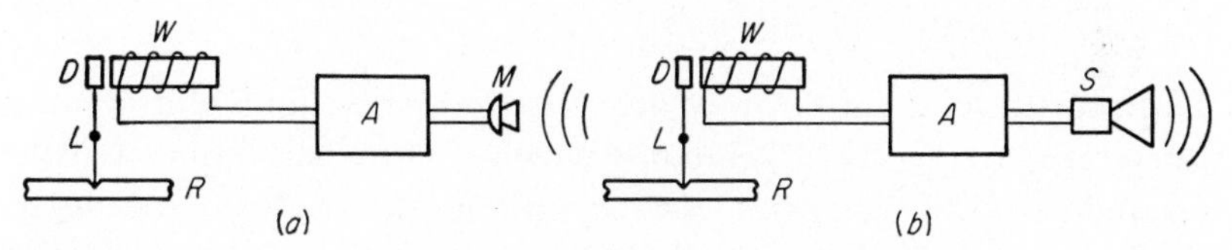

FIG. 16–3. Diagrammatic sketch of magnetic recording and reproducing assemblies.

magnetic pickup.) The minute current developed by the pickup device is strengthened by means of the amplifier, *A*. The amplified current, in turn, serves to actuate the loudspeaker, *S*.

A so-called **crystal pickup** is sometimes employed. In this unit the upper end of the pivoted needle is in contact with a certain type of crystal which, when subjected to a variable pressure, develops a corresponding electrical potential (piezoelectric effect). This variable potential establishes an alternating current which in turn is amplified as in the case of the magnetic unit.

While the electromechanical technique is a decided improvement over the older mechanical method of recording and reproducing, it has rather distinct frequency and other limitations. Owing to the mechanical structure of the cutting and reproducing heads it is not possible, by the process above described, to record faithfully below 100 or above approximately 6000 cps. Further, the relative intensity of all harmonics is not maintained, and this means that real high fidelity is not possible.

A second limitation has to do with dynamic range. If the sound being recorded is loud the amplitude of swing of the cutting stylus will cut into the preceding groove. This can be avoided by increasing the spacing between grooves, but to do so decreases the amount of recording that can be made on a single disk. One remedy, sometimes applied in such cases, consists in cutting down the amplification of the louder passages when recording and then making use of a playback amplifier so designed that it will restore the louder passages to their original relative sound level. The so-called long-playing recordings have, to some extent, remedied the above-mentioned limitations, but there is much still to be desired when it comes to fidelity.

16–2. *Wire and Tape Recording*

In recent years an entirely different recording and reproducing procedure has come into relatively wide use. This technique involves a somewhat different principle than the methods above described. Many years ago V. Poulsen, a Danish scientist, developed a method by which sound waves could be made to produce a magnetic record on a moving steel wire. The essentials of the Poulsen assembly are sketched in Fig. 16–4. By means of a suitable driving mechanism, a thin steel wire or tape, W, was caused to unwind from one spool onto another. The recording head, H, consisted of a pointed permanent magnet surrounded by a winding which was connected in series with a battery, B, and a microphone, M. As sound waves impinged upon the microphone the resulting variable current modified the strength of the magnetic field, thereby giving rise to areas of varying magnetic intensity in the moving steel wire. Thus there was produced a permanent magnetic record of the incident sound waves.

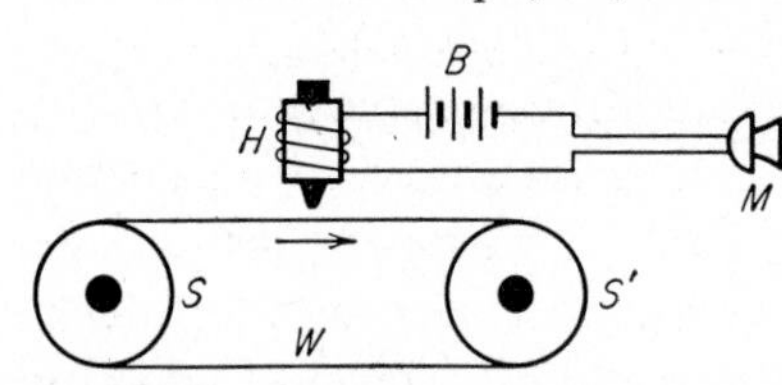

FIG. 16–4. Essentials of wire-recording assembly.

In order to reconvert this magnetic recording into sound waves the microphone and the local battery were replaced by a telephone receiver. If now the wire is wound back to the beginning of the record and again caused to move forward in the original direction at its original speed the magnetic "spots" in the wire will produce, by electromagnetic induction, a varying current in the winding of the electromagnet, H; and this current, acting through the agency of the telephone receiver, will reproduce the original sound.

Poulsen's apparatus was originated long before amplifiers were known. In the modern wire recorder the original microphone current is amplified, as is also the reproduced current; and the head receiver is replaced by a loudspeaker. The wire travels at a rate of 3 to 15 in./sec, and a spool may carry enough wire for a sixty-minute recording.

The above-mentioned wire-recording procedure has recently been more or less replaced by a method that makes use of a so-called magnetic tape. Such a tape is made of paper or some plastic material, is about ¼ in. (6 mm) wide, and on one of its sides there is deposited a very thin layer of finely divided magnetic material. The general principles involved in the use of such a magnetic tape in the recording and reproducing processes are the same as in the wire-recording system. Magnetic sound records may be stored for a considerable period of time. If desired, a record may be erased by passing the wire or tape through a magnetic field, thus making it possible to use the same medium repeatedly. The tape is more convenient to use mechanically than the wire, and its acoustical characteristics are somewhat better. The development of the modern tape recorder constitutes a remarkable engineering achievement.

16–3. *Microphones*

In order to make a record of sound waves some means must be available whereby the harmonically varying pressure of sound waves may be converted into a varying electric current. A device that serves to bring about energy transformations from one form to another is known as a **transducer;** a microphone is an agent of this nature.

So far as our purpose is concerned the term microphone is a misnomer. Strictly speaking, a microphone, as used in ordinary telephone communication, applies to a device in which the active components consist of a small quantity of granular carbon held between two polished carbon plates. Carbon has a property by virtue of which its resistance to the passage of the electric current changes with the pressure to which it is subjected. This is spoken of as microphone action. While such a device is quite satisfactory for ordinary telephone use, and as a component in mobile radio units, it is not suited to serve as a transducer in connection with sound-recording and sound-amplifying systems—it is more or less noisy and its frequency limits are rather narrow. Because it was the first device used as a sound transducer, the term microphone is now commonly applied to any device that is capable of transforming the energy of sound waves into electrical energy.

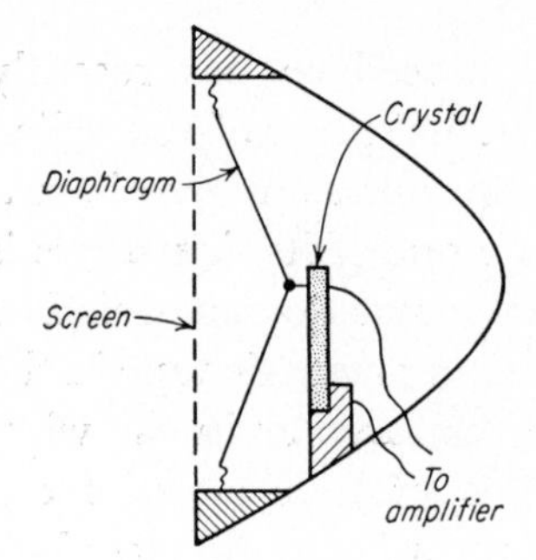

Fig. 16–5. Essentials of crystal microphone.

There are, in general, two types of modern sound transducers (microphones): One is actuated by the pressure of sound waves, while the other responds to the particle velocity due to the passage of the sound waves through the air. The former is designated as a pressure microphone and the latter as a velocity microphone. The original carbon unit is a pressure-actuated device. There are, as we shall see, two forms of the pressure-type units.

We have previously referred to the fact that certain crystalline substances will, when subjected to pressure, develop an electrical potential difference between different parts of the crystalline structure—the piezoelectric effect. A widely used type of sound pickup (microphone) makes use of this unique property possessed by certain crystals. Rochelle salt exhibits this property to a marked degree, and is widely utilized as a sound responsive agent in what are commonly known as **crystal microphones.** Figure 16–5 shows the essential components of a crystal pickup. A diaphragm is mechanically articulated with one surface of a suitably mounted rochelle salt crystal, as shown. Electrical connections are made to two points on the crystal, and these connections feed into an amplifier. The pressure of sound waves as they impinge on the diaphragm causes it to vibrate and exert a varying pressure on the sensitive crystal. The output from the crystal will be an electrical replica of the sound-wave pressure. Crystal microphones are simple devices, and hence relatively inexpensive. They are used in many low priced public-address systems. However, this type of pickup has certain inherent mechanical and frequency limitations that make it unsuitable for high-grade recording and public-address work.

Another pressure-operated unit, known as an **electrodynamic microphone,** is a superior type of sound transducer.

A well-known principle in elementary physics states that if and when a closed electrical conductor is caused to move in a magnetic field, an electric current will be set up in the moving conductor—an *induced* current is established in the wire.

This principle is utilized in the design of this type of transducer. Referring to the sketch shown in Fig. 16–6, we see a permanent magnet, a soft iron core, and a coil of wire free to move in the space between the poles of the magnet and the iron core. This voice coil, as it is called, is mechanically attached to the diaphragm.

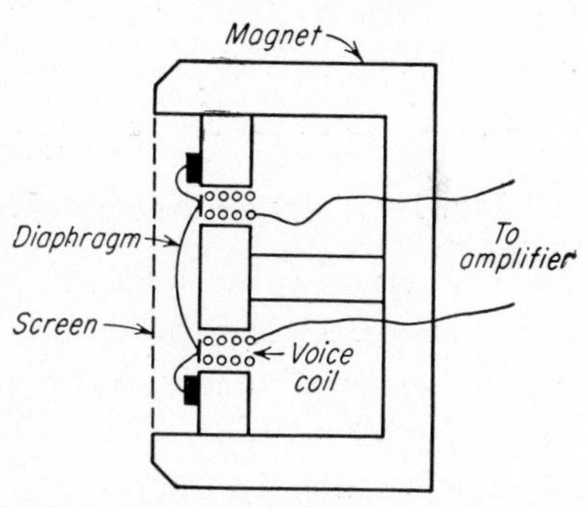

Fig. 16–6. Essentials of dynamic microphone.

Wires lead from the terminals of the voice-coil winding to an amplifier. The vibration of the diaphragm, caused by the incident sound waves, results in a periodic motion of the voice coil in the magnetic field. Thus the energy of the sound waves is transformed into electrical energy, which, when properly amplified, will serve to actuate a recording head or a loudspeaker. When the unit is properly designed the electrical output of such a transducer is a replica of the sound input, and it will respond to frequencies between 40 and 12,000 cps. The frequency response is, however, not flat (free from peaks). This is evident from an examination of the response curve (Fig. 16–7) of one of the better units of this type.

A diagram of a **velocity microphone** is shown in Fig. 16–8. In this assembly the voice coil of the previously described type of unit is replaced by a metallic ribbon supported, under light tension, between the poles of a permanent magnet. This ribbon is caused to move by the difference in sound pressure on its two flat sides. This differential pressure is due to the difference in particle velocity at these two points. The movement of the ribbon develops an output voltage as in the dynamic unit. The electroacoustical characteristics of the velocity type of pickup are much the same as in the case of the dynamic transducers, with one exception. Whereas the trans-

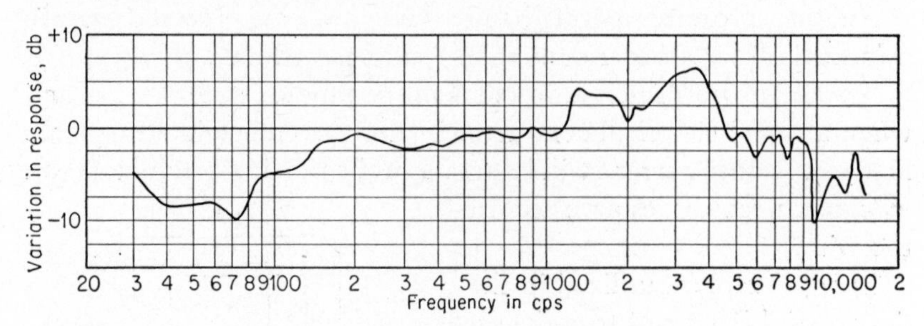

FIG. 16–7. Response curve of high-grade dynamic microphone.

ducer previously described will respond to the sound waves arriving from any direction, the ribbon microphone, as it is called, will respond to sound waves that reach the unit from the front and the rear only. This results in a pickup pattern such as that shown in Fig. 16–9*a*; such a response is referred to as **bidirectional.**

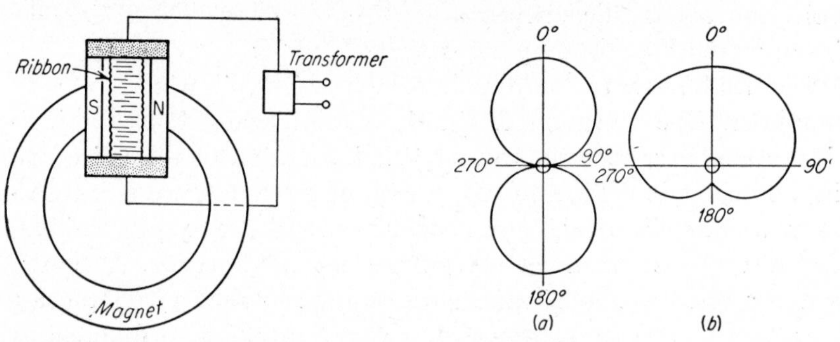

FIG. 16–8 (*left*). Diagram of velocity microphone.
FIG. 16–9 (*right*). Response curve (*a*) of bidirectional microphone, and (*b*) cardioid type unit.

Without going into technical details, it may be said that any of the microphones already described may, by proper acoustical design, be made to give a **unidirectional** response, such as indicated in Fig. 16–9*b*. Such a directional response is commonly referred to as a **cardioid** pattern. Since a microphone having a cardioid response pattern responds only to sound waves that arrive through a fairly narrow front angle, its use largely eliminates the effect of sounds coming from the rear. It is because of this important characteristic that unidirectional microphones are used almost universally in studio recording and in connection with sound-reinforcing systems.

Reference has been made to the fact that the microphones thus far described give a response that is not free from frequency peaks—frequency distortion. There is one type of sound transducer that does show a response graph that is practically flat. Reference is made to the **condenser microphone.** This is a very simple device, and makes use of the variable capacitance principle that we have already discussed in connection with one form of the electronic organ. The essential components of a condenser type of transmitter are shown in Fig. 16–10. A thin metal diaphragm is located extremely near to, and parallel with, an insulated metal plate. These two members constitute a condenser. Provision is made to apply a potential difference of some 200 volts between the diaphragm and the plate. Sound waves impinging on the diaphragm cause it to vibrate and thus change the capacitance of this elemental condenser. As in the organ case, any change in capacitance brings about a change in the strength of the charging current, and this in turn develops a corresponding potential difference at the terminals of the resistor R. The electrical output of such a unit is quite small, but the level can be raised to the desired value by means of suitable amplification. The condenser microphone is **omnidirectional** in its pickup pattern, but gives a fidelity of response that is unequaled by any other types of sound transducer. The response of such a unit is indicated in Fig. 16–11.

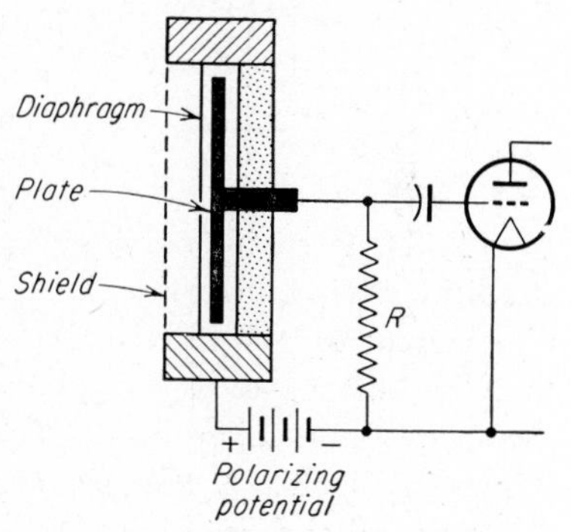

FIG. 16–10. Sketch of condenser microphone, and associated amplifier.

In the design of all types of microphones special precautions are necessary to avoid resonance effects in connection with the vibratile member. If the natural frequency of the diaphragm should chance to fall within the range of frequencies being used, resonance effects would tend to distort the response. Various constructional measures are employed to minimize such effects.

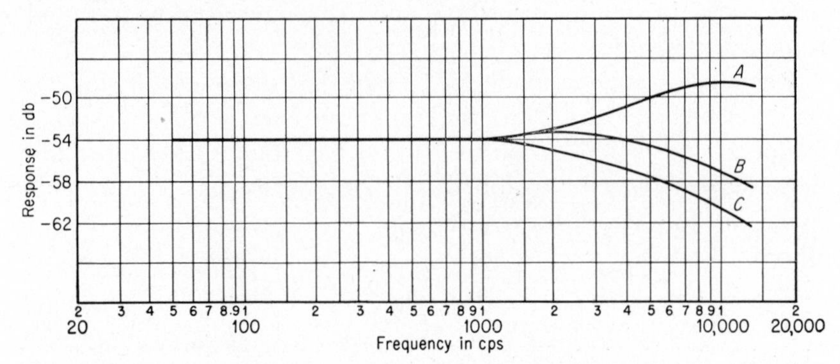

Fig. 16–11. Response curve of condenser microphone under the following conditions: *A*, with grille in place, sound incidence perpendicular to diaphragm; *B*, with grille, incidence parallel; *C*, without grille. Note that the best response occurs when sound waves are moving parallel to the face of the microphone. Compare these response curves with that shown in Fig. 16–7. (Graphs reproduced by permission of the Kellogg Switchboard and Supply Co.)

16–4. *Amplifiers*

If high-fidelity recording and sound amplification is to be secured there is needed, in addition to a first class microphone, an amplifying system whose electrical output will be identical in waveform to that which is passed into the unit. The term "high fidelity" means just that, but there are few commercial amplifiers that can be depended upon to meet such a specification. It is beyond the scope of this volume to enter into a detailed discussion of amplifier design. It must suffice to point out certain criteria that should guide one in the selection or construction of a high-grade amplifying assembly. These criteria are:

1. The amplifier should have a rated output at least 50 per cent greater than will probably be required. This is because most amplifiers show serious distortions when operated at their rated output. Never overload an amplifier.

2. The amplifier should be absolutely free from hum or other background noise. This is difficult, if not impossible, to attain unless the "power pack" is constructed as a separate unit. Another factor in securing quiet operation is to employ a low-impedance microphone (or phonograph pickup) and a low-impedance well-shielded input transformer.

3. The last two stages of the assembly should be of the push-pull type, and transformer coupled.

4. High-grade audio transformers should be used throughout.

5. A test should be run to determine whether distortion is present. Such a test can be carried out by feeding into the amplifier a sine or a square wave and examining the output waveform by means of an oscilloscope when the amplifier is connected to its normal load. A further test of fidelity consists in activating the amplifier at various frequencies by means of an audio oscillator and noting the voltage at the output terminals under load conditions. Technically speaking, a good amplifier should not show a frequency deviation of more than ± 0.5 db from 30 to 15,000 cps; not more than 1 per cent harmonic distortion at full load; and intermodulation not to exceed 1 per cent at full output.

Any trustworthy dealer in amplifiers should be able to state whether the unit complies with the above-listed specifications, and should be equipped to carry out the tests above suggested, in support of his claims.

16–5. *Loudspeakers*

Assuming that one has a reasonably good reproducing head, or microphone, and a correspondingly satisfactory amplifier, there still remains one other possible cause of defective response. We refer to the transducer commonly known as a loudspeaker—the unit that serves to reconvert the electric current into sound waves.[1]

A loudspeaker is essentially the same as an electrodynamic type of microphone, but operating in the reverse order. In the case of the loudspeaker both the diaphragm and the voice coil are, in

[1] The remarks that follow regarding the performance of loudspeakers apply also to the performance of transducers that form a part of radio sets and sound equipment in general.

general, larger than those to be found in the microphone. But the basic principle of operation is the same. In the microphone, sound waves serve to develop an electric current. In the loudspeaker the alternating current from the amplifier is fed into the voice coil, thus establishing a varying magnetic field within that coil. This variable field reacts with the magnetic field due to the permanent magnet and thereby gives rise to a force that moves the diaphragm harmonically, which in turn sets up corresponding sound waves. Thus through the medium of such a transducer electrical energy is converted into the energy of sound waves—the conversion cycle being thereby completed. That the speaker-microphone combination is a reversible process is evidenced by the fact that a speaker may be, and often is, utilized as a sound pickup.

In general, it may be said that the average loudspeaker is the least satisfactory of all the components which go to make up a recording-reproducing system. There are three principal reasons for this situation:

1. The small size (diameter) of most speakers
2. The fact that many speakers are built to a competitive price
3. The inherent technical difficulties involved in the design and manufacture of a speaker having a wide dynamic and frequency range

Speakers whose diameters are less than 12 in. show a marked drop in response in the frequency band covered by the lowest octave and a half on the piano. Indeed, there is only a feeble bass response from the speakers commonly incorporated in portable and small table model phonographs and radio sets. Fortunately, owing to the formation of subjective tones (Sec. 7–5), the ear may supply, in part at least, the missing base tones. Nevertheless, the final result is acoustically imperfect and far from satisfactory to the listener who has any appreciation of musical quality. Speakers whose diameter is of the order of 18 in., when associated with a suitable acoustical baffle, will yield base tones which bear an approximate resemblance to the original sound. It is, however, extremely difficult, if not impossible, to design a one-unit speaker which will reproduce both the low and the high frequencies at their true relative values. The response of a single unit speaker is shown in

Fig. 16–12. Note the wide variation in the response at different frequencies. When it is recalled that the ear can easily detect a change of 2 db in sound level it is obvious that, under the circumstances indicated, high fidelity is impossible. To avoid this difficulty it is the practice to combine two speakers in a single reproducing unit. In the two-unit model one speaker is designed to give a fairly flat

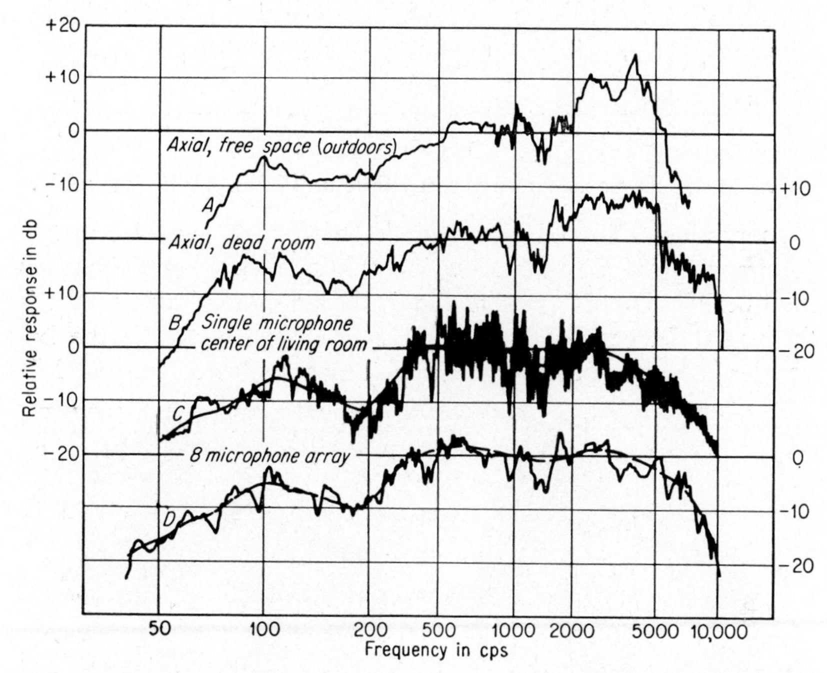

FIG. 16–12. Comparative response curves for a given single-unit speaker measured under various conditions. (From *Technical Monograph No.* 1, Jensen Mfg. Co.)

response up to, say, 1200 cycles. The associated high-frequency unit will respond well from slightly below 1200 up to 10,000 or 12,000. Such a dual reproducing unit gives a much more satisfactory response than any single speaker. Recently a three-unit assembly has appeared on the market. Figure 16–13 shows the response which may be attained when a multiple-speaker unit is employed. Note the marked improvement over the single unit assembly. In Fig. 16–14 may be seen a single unit speaker and also one having the conventional low-frequency cone ("woofer") plus a coaxially mounted high-frequency unit ("tweeter"). The coaxial units have a fairly wide angle of radiation for both the high and the low frequencies. If a concentrated sound beam is desired a so-called reentrant horn is commonly employed, such as shown in Fig. 16–15. This type is used in outdoor installations.

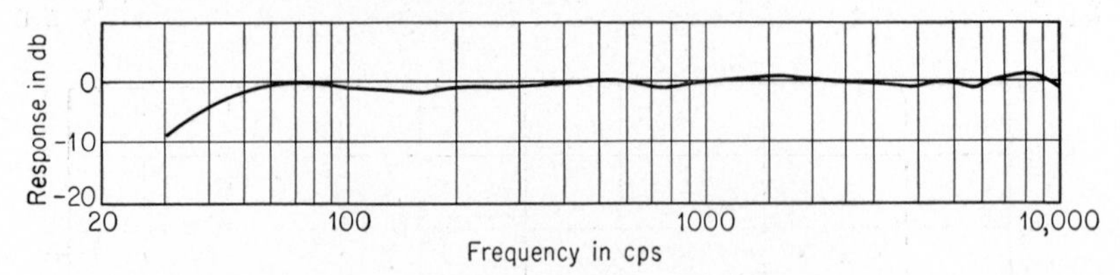

FIG. 16–13. Response curve of a multiple-unit loudspreaker assembly.

In passing it should be noted that there are several **external** factors which modify the acoustical results to be secured from any loudspeaker. One of these factors is the position of the speaker in the room. In general, it may be said that a corner is the best position. If the speaker assembly is not enclosed, it should be

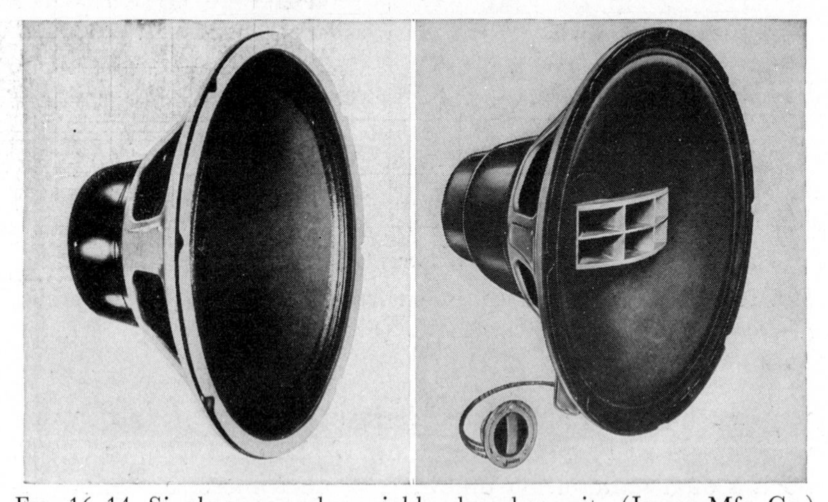

FIG. 16–14. Single cone and coaxial loudspeaker units. (Jensen Mfg. Co.)

placed directly against a flat wall; and the speaker opening should be at ear level, not near the floor.

A second factor which tends to modify the acoustical results is the position of the listener with respect to the sound axis. This is shown by the automatically traced response curves appearing in

Fig. 16–16. The graphs show that below about 200 cycles there is no appreciable effect so far as the frequency-response relation is concerned. If, however, the listener is located off the axis the sound level of the higher frequencies is lower than on the axis. At a

FIG. 16–15. Reflex trumpet type of loudspeaker. (University Loudspeakers, Inc.)

60° position the effect is even more pronounced, and at a point 90° off the axis the "highs" fall to a relatively low value. At 2000 cycles, for instance, the response is down 18 db when compared to the sound level on the axis of the speaker. Such a drop in the

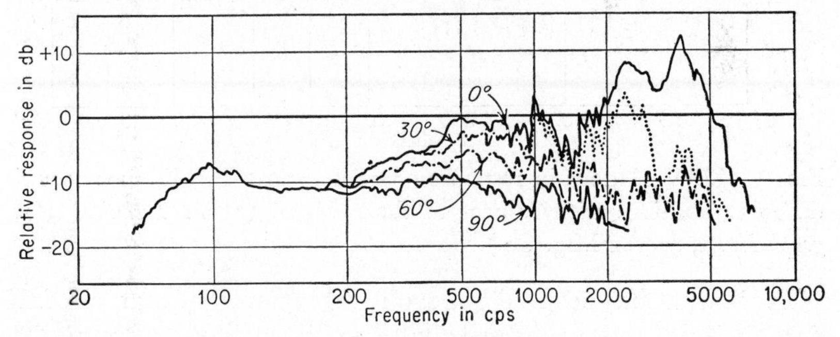

FIG. 16–16. Outdoor response curves at various distances off the sound axis. (*Technical Monograph No.* 1, Jensen Mfg. Co.)

value of the highs results in a marked diminution in the intelligibility of speech and in a decided loss in the fidelity of musical tones.

Even when the reproducing assembly and the listener are indoors the relative position of the auditor constitutes a factor in the over-all acoustical results. In Fig. 16–17 are to be seen the response curves for several listener positions. The reflection from the walls, ceiling, and floor tends to reduce somewhat the directional characteristics which are observed when listening out of doors; but even when listening inside the position of the auditor has an appreciable effect on the net results.

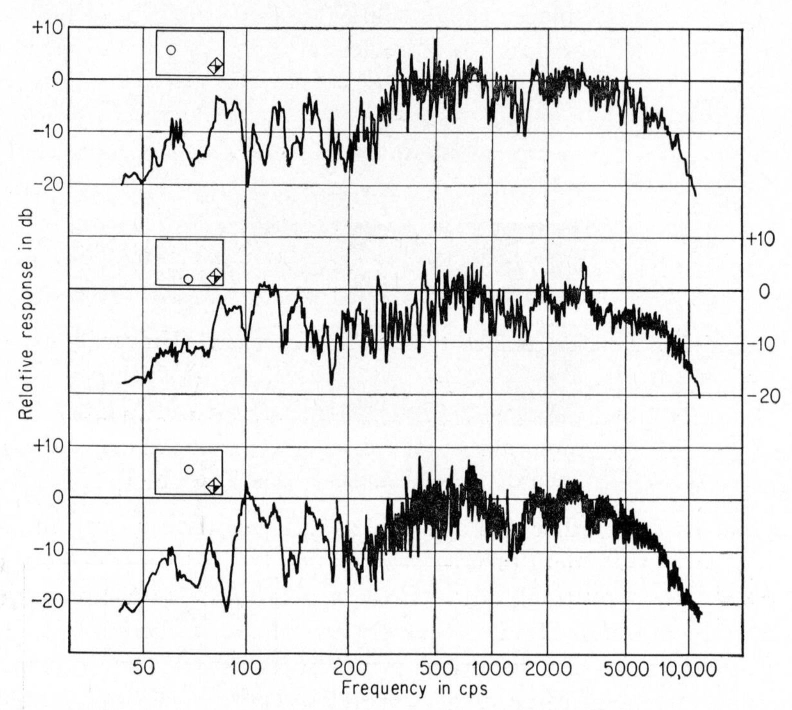

FIG. 16–17. Variations in response characteristics when listener occupies different positions with respect to the speaker. (From *Technical Monograph No.* 1, Jensen Mfg. Co.)

From what has been said above it will be evident that, as of this date, it is not to be expected that a small single-unit loudspeaker, placed in almost any position in a room, will give good acoustical results.

In concluding our discussion of microphones, amplifiers, and loudspeakers, it remains to point out that a sound-recording or -reproducing system, or a public-address layout, should be planned as a unit. If good over-all results are to be attained the electrical and acoustical characteristics of the several components should match. A collection of unmatched elements will not give satisfactory results. The assembly should be engineered as a whole.

16–6. *Film Recording and Reproduction*

The important place which talking motion pictures occupy in present-day life warrants a brief review of the techniques employed in the recording and reproducing of voices and musical sounds on photographic films.

There are two principal methods by which sound is currently recorded photographically. One is referred to as the variable-area

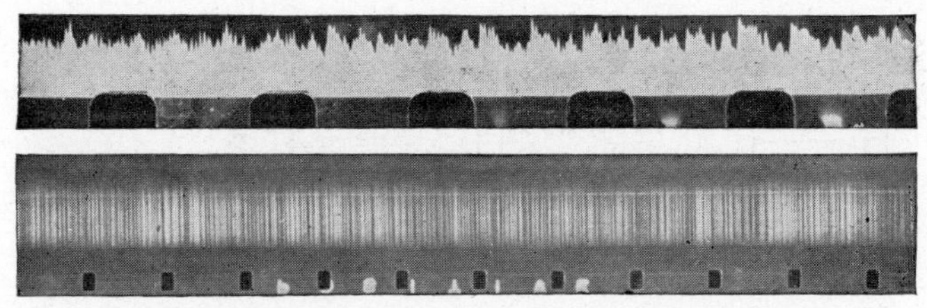

FIG. 16–18. Photographically produced sound records ("sound tracks"): *upper*, variable-area record; *lower*, variable-density recording.

method, and the other as the variable-density method. Figure 16–18 shows an enlargement of *sound tracks* made by these two processes. The actual sound track on a 35-mm film is about 2 mm in width, and on the 16-mm film it is scarcely half of that width.

Any sound-on-film recording procedure involves a microphone, an amplifier, and some device whereby the intensity of the light on the sound-track area may be modulated by the output of the microphone-amplifier train of sound equipment. In the case of variable-density recording there are several procedures by which this type of record may be made. One such method involves a so-called light valve by which the intensity of the light from a small incandescent lamp is caused to vary in conformity with the amplifier output. Another method involves a "glow lamp," the output of which can be modulated by the output of the sound amplifier. The essentials of the latter system of recording are sketched in Fig. 16–19. The output of a sound pickup, M, is amplified and connected through a transformer, T, to the glow lamp, G. This glow lamp is similar to the small neon lamps in common use. Such a lamp consists of a glass enclosure equipped with two electrodes and containing a gas at low pressure. When the voltage of the local battery, *B*, is raised to a certain value, current will pass between the two electrodes in the form of a luminous discharge, the color of the emitted light depending upon the nature of the residual gas. For sound recording, a gas giving a white or bluish glow is employed. If the voltage of the local battery is adjusted so that a discharge takes place, and if by any means the voltage applied to the tube is changed in value (increased or decreased) the intensity of the electrical discharge will change in a corresponding manner. As the electrical discharge changes, the intensity of the light also will

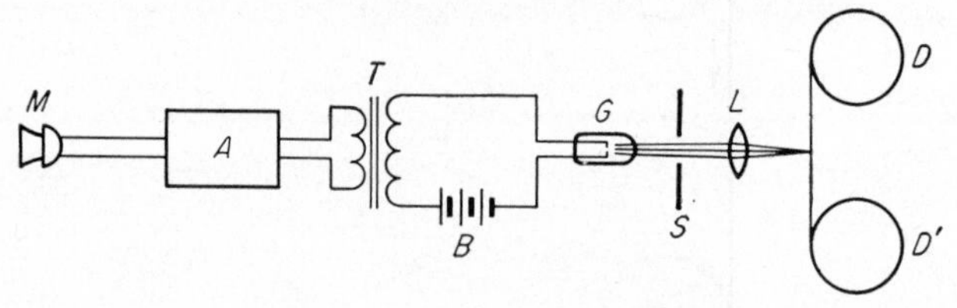

FIG. 16–19. One of several variable-density schemes used in sound recording.

change, thus bringing about a **modulation** of the light beam. The alternating voltage supplied by the voice-actuated amplifier serves to bring about the necessary variations in the voltage applied to the lamp, and **hence the sound waves will act to modulate the intensity of the light issuing from the lamp.**

The modulated light beam, after passing through a narrow slit, is brought to a focus on the moving photographic film. By this means a photographic record is made of the modulated light beam. The fundamental and the overtones of a given sound are thus faithfully recorded as photographic lines of varying density, as seen in the lower record of Fig. 16–18.

In producing a variable-area record the recording mechanism involves the use of a so-called light gate, actuated by the input amplifier. The gate serves to modify the length of the slit exposed to the exciter lamp, thus resulting in the "hill-and-dale" type of record indicated in Fig. 16–18. Both forms of record may be reproduced by the same type of equipment, and they are in fact used interchangeably. The variable-density method is used for certain types of recording while the variable-area scheme is employed in other cases. Recently magnetic sound-on-film recording has been used with some success. The principles involved in this technique are the same as those made use of in magnetic-tape recording of sound, previously discussed.

When using an optically recorded sound record the transformation of such a record into sound waves is accomplished through the agency of a photoelectric cell. This is an electronic device by means of which light causes the liberation of electrons from a

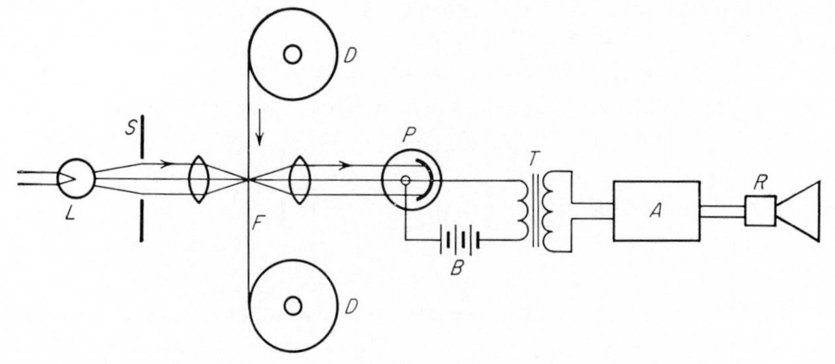

FIG. 16–20. General plan of sound reproduction from photographic type of recording.

specially prepared metallic surface. These liberated electrons, as they are caused to move along a wire under the influence of a source of electrical potential, constitute an electric current. This feeble current may be amplified and caused to operate a loudspeaker. The relation of the essential components required to convert the photographic sound record into audible sound is indicated in Fig. 16–20. Light from a so-called exciter lamp *L* (similar to the bulb in a common flashlight) passes through a narrow slit, *S*, after which it is focused on the sound track of the moving film, *F*. A second lens changes the light rays into a parallel beam which in turn enters the photoelectric cell, *P*, and thereby sets up an electric current in the cell circuit. The photographic lines constituting the sound track cause a variation in the intensity of the light incident on the light-sensitive surface of the photoelectric cell. The resulting variable current is fed into an amplifier through the transformer,

T, or its equivalent. The electrical output of the amplifier actuates the loudspeaker, R. Thus the original sound is reproduced and, everything considered, with amazing fidelity and greatly augmented volume.

Recently a new type of electronic tube has been developed which functions as a combined photoelectric cell and high-gain amplifier. The use of this new unit will probably result in higher fidelity and make possible a more compact reproducing assembly. It is not too much to expect that complete major operatic productions may soon be recorded and reproduced with a high degree of tonal fidelity. Indeed much has already been accomplished along this line. Here we see another striking example of the way in which science is making it possible for large numbers of people to hear and enjoy the world's greatest musical compositions.

17 *Architectural Acoustics*

17–1. *Reverberation*

Since music is rendered, and words are spoken, to be heard it is obviously important that the acoustical environment be such that the auditor may be able to hear the musical composition or the spoken message in a satisfactory manner. The acoustical characteristics of public gathering places therefore become important. For this reason we shall consider, briefly, certain aspects of architectural acoustics.

In Sec. 3–6 the subject of the reflection of sound was discussed, and it was there pointed out that any sound originating in a room may give rise to a series of multiple reflections. In an enclosure such as a classroom, a concert hall, or an auditorium, a given sound may indeed undergo several hundred reflections before the wave energy is entirely dissipated. Because of the short distances involved these multitudinous echoes may blend into what amounts to a prolongation of the original sound that may persist, with decreasing intensity, for several seconds after the source has ceased to radiate. This prolongation of the original sound, due to many reflections, is designated by the term **reverberation.** Within certain limits this "building up" of the sound in a room may serve a useful purpose; but, on the other hand, if unduly prolonged it may seriously interfere with the intelligibility of speech and the proper reception of music.

The growth and decay of sound in a reverberant room may be graphically represented as shown in Fig. 17–1, which should be carefully studied. If the sound emitted by the source be prolonged over a period of several seconds, a steady acoustical condition will at length come to exist. The "growth" aspect of the process is indicated by the curved part of the graph between O and X: the steady state condition obtains when the radiated energy is being absorbed, or otherwise dissipated, at the same rate at which it is emitted by the source. The length of time required to bring about this condition of equilibrium, represented by Ot on the time axis, will depend upon the size of the room and upon other acoustical factors. Time is required to set the whole body of air into vibration.

If at any instant, represented by t' on the diagram, the source ceases to emit energy, the sound in the room will not immediately

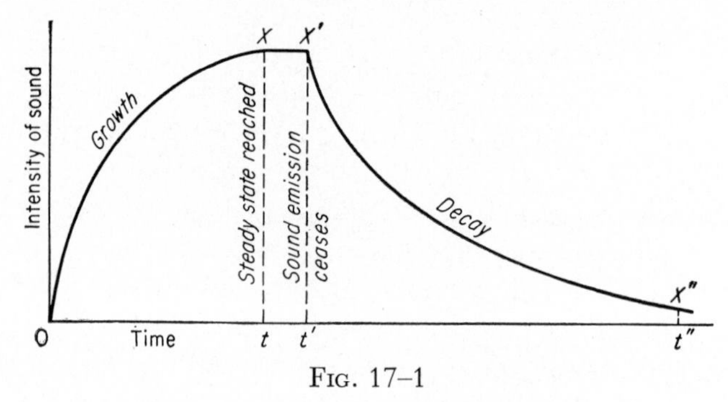

Fig. 17–1

cease but will gradually diminish in intensity as indicated by the curve $X'X''$.* The length of time required for the sound to die out (become inaudible) will depend upon the size of the room and also upon the absorbing effectiveness of the interior surfaces of the room. This time interval of decay, represented by $t't''$, is known as the **reverberation time,** or **time constant.** Quantitatively this reverberation time of a room is defined as the time required for the sound of a given frequency to decrease to one millionth of its original intensity. This would be represented by $t't''$ on the time axis of the chart in Fig. 17–1.

It should be added in passing that the smooth decay curve

* To those who are familiar with mathematical concepts both parts of the graph will be recognized as being logarithmic in character. The equations corresponding to the growth and decay curves are to be found in V. O. Knudsen, "Architectural Acoustics," 1932 ed., p. 129, and also in W. Y. Colly, "Sound Waves and Acoustics," p. 211.

shown in Fig. 17–1 does not tell the complete story. Actually the phenomenon of reverberation in a room is more or less irregular, and takes on some such a decay form as that indicated in Fig. 17–2. The smooth curve shown in Fig. 17–1 would roughly represent

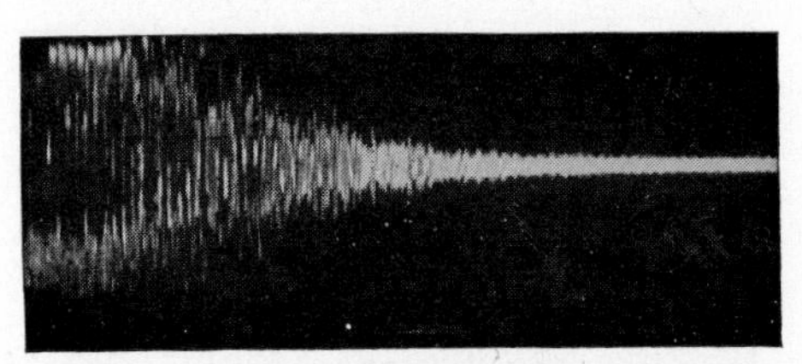

Fig. 17–2. Representative oscillographic record of the decay of reverberation in a room. (Acoustical Materials Association.)

the envelope of the individual sound-wave vibrations recorded in Fig. 17–2.

The time interval of decay is of great importance. If this constant is too great, as is often the case, speech becomes an unintelligible jumble of sounds, and music loses its true character. The reason for such an unfortunate condition becomes apparent when one examines the condition by means of a graphic representation, as shown in Fig. 17–3. Suppose, for instance, that sound is emitted intermittently, and suppose that the interval between sounds is less than the time of decay (reverberation time). In the diagram (Fig. 17–3) the boundary of each of the shaded areas may, for example, represent the growth and decay of sounds involved in successive spoken words. In the illustration the steady state is not reached; but, because of room reverberation, the sound of any given word persists while the following word is spoken. As a result of this overlapping an auditor may not be able to understand the speaker. Thus we see that while the persistence of acoustical energy

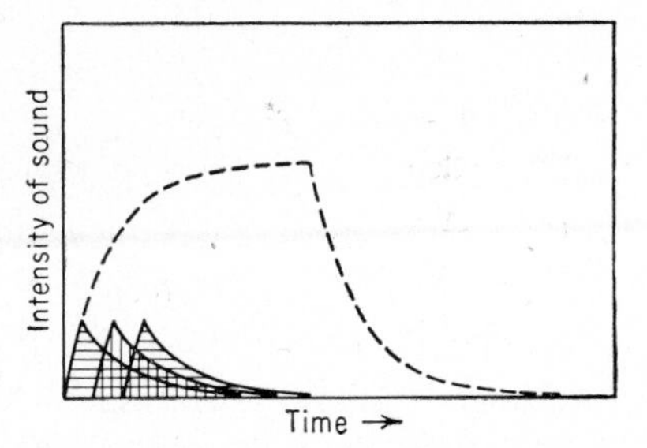

Fig. 17–3. (Reprinted by permission from "Acoustics of Buildings" by F. R. Watson, published by John Wiley & Sons, Inc.)

may assist a speaker to "fill" the room, it may, if the continuance is too long, bring about marked confusion; **loudness serves no purpose unless understandability also obtains.**

There are, of course, factors other than reverberation that affect the intelligibility of speech. Among these are the shape of the room, the loudness of the spoken words, and the masking effects of extraneous sounds. What are called **"percentage articulation"** tests of a given auditorium are sometimes made. In carrying out such a test a speaker stands in the normal position and repeats a large number of meaningless syllables or phrases. These are listened to by several persons stationed at various positions in the auditorium. Each auditor makes a written record of what he thinks he hears. The average percentage of the sound that is heard correctly by the listeners is the percentage articulation of that particular room. A result of 85 per cent indicates reasonably good acoustical conditions; 75 per cent fairly satisfactory; 65 per cent just passable; while less than 65 per cent is considered to be entirely unsatisfactory. Such a test is not difficult to make and the results will give a useful index of the acoustics of the room, so far as speaking is concerned. In the event that the acoustical conditions are found to be below par, steps can be taken to modify one or more of the controlling factors. The control of reverberation is discussed in the next section.

In the case of music the overlapping of successive sounds is not as objectionable as in the case of speech; it produces the same effect as holding down the sustaining pedal on the piano. However, if the decay occupies too long an interval of time, the results may be unsatisfactory even in the rendition of music. Figure 17–4 diagrammatically represents the situation in the case of music.

We are, then, faced with the question as to what is the **optimal** decay time—a period sufficiently long to take advantage of the reinforcement of the primary sound, and yet not long enough to cause serious acoustical confusion.

The late Professor Sabine, of Harvard University, laid the foundation of modern architectural acoustics. His pioneer work cleared away many of the traditional misconceptions concerning behavior of sound in auditoriums.[1] As a result of the research work originally done by Professor Sabine, and the subsequent

[1] See W. C. Sabine, "Collected Papers in Acoustics."

important investigations by Watson, Knudsen, and others, it has been established that the optimal reverberation time of an enclosed space depends upon the volume of the room and also upon the use to which the room is to be put. Acousticians are now more or less

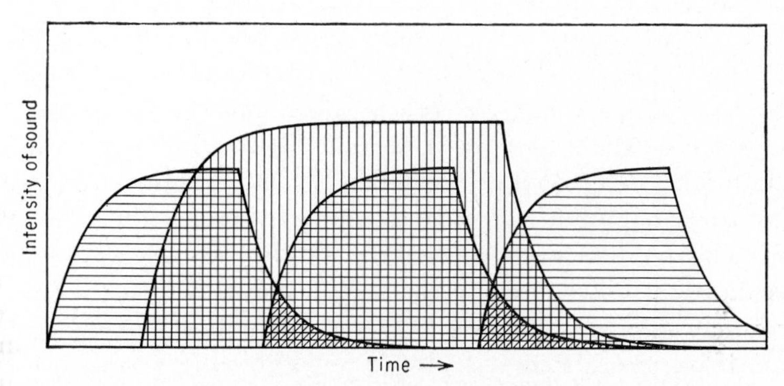

FIG. 17–4. (Reprinted by permission from "Acoustics of Buildings" by F. R. Watson, published by John Wiley & Sons, Inc.)

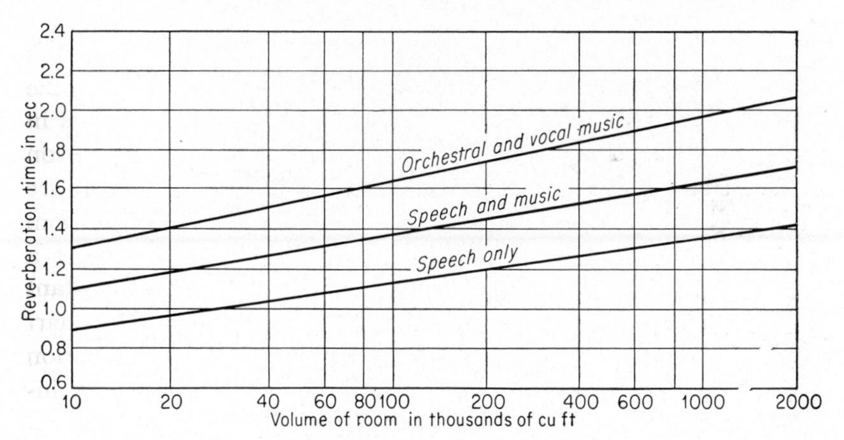

FIG. 17–5. Optimum reverberation time at 500 cycles, as a function of room volume.

in agreement as to the value of the time constant that should obtain under a given set of circumstances. The chart shown as Fig. 17–5 embodies the integrated study and experience of several leading workers in the field of applied acoustics.

As previously indicated, the optimal reverberation time is a function of the frequency involved, having a somewhat higher value for frequencies below 500 cycles than for frequencies above that value. However, in practice, computations are usually based on a frequency of 500 cycles.

17–2. *Control of Reverberation*

Every room has a natural reverberation time. If one knows the dimensions of the room, it is possible to determine by simple computation what that time constant is. Professor Sabine developed an empirical equation giving the natural reverberation time of any room. This relation takes the form $t = 0.05V/A$, where A represents the total absorption by the surfaces upon which the sound waves may impinge.[1] The term A is a composite factor made up of a series of terms, thus

$$A = a_1s_1 + a_2s_2 + a_3s_3, \text{ etc.}$$

where a_1, a_2, a_3, etc., represent the absorption coefficients of the corresponding areas s_1, s_2, s_3, etc. In the above formulae V is in cubic feet and s is square feet. The term as is expressed in absorption units, now called sabins after Professor Sabine. Knowing the nature of the surface material of the walls, ceiling, floor, etc., of a room interior, we can thus compute the reverberation constant of a particular interior. If such a computation shows the time constant to be seriously in excess of the optimal value, as given by the chart in Fig. 17–5, it is easy to determine what additional absorption must be provided. A simple example will illustrate the steps involved in such a determination.

Let us first compute the reverberation constant of a room for which no attempt has been made to control this factor. Suppose we have a room that is to be used for both speech and music with the following specifications:

[1] Several other formulas by which the reverberation time can be computed have been developed from theoretical considerations. See R. F. Watson, "Acoustics of Buildings," 3d ed., p. 33; or L. L. Beranek, "Acoustical Measurements," p. 862. The Sabine equation is, however, sufficiently accurate for the purposes of this book.

Floor, hard wood, 25 × 50 ft
Walls, plaster, 15 ft
10 windows, each, 2.5 × 6 ft
125 seats, plain, covering 90% of floor
Computed volume, 18,750 cu ft

The total absorption may be computed as follows:

Walls and ceiling, less windows	3350 sq ft × 0.04	= 134	sabins
Windows	150 sq ft × 0.027	= 4	sabins
Floor, less seat coverage	125 sq ft × 0.03	= 4	sabins
Seats	125 × 0.1	= 12.5	sabins
Total		154.5	sabins
One-third audience	41 × 4	= 164	sabins
Maximum audience	125 × 4	= 500	sabins

t (no audience) = 0.05 × 18,750/154.5 = 6.1 sec
t (⅓ audience) = 0.05 × 18,750/318.5 = 3.0 sec
t (maximum audience) = 0.05 × 18,750/654.5 = 1.4 sec

Referring to the room graph given in Fig. 17–5, it will be seen that the optimal reverberation time for a room of the size indicated is about 1.15 sec. Our computation shows that, even with the room filled, the time constant is somewhat too high and when only one-third filled the reverberation time is far too great. What can be done to remedy this situation? Obviously a certain portion of the walls or ceiling will have to be covered with sound-absorbing material. The next question is: How much surface must be so treated? Fortunately we can easily arrive at the answer.

Taking 1.15 sec as the optimal reverberation time and using the Sabine formula we can compute the *total* absorption required for the room in question. Substitution yields

$$A = 0.05 \times \frac{18{,}750}{1.15} = 815 \text{ sabins}$$

From our original computation we see that, say for one-third audience, the absorption is 318.5. Now

$$815 - 318.5 = 497.5$$

It is therefore evident that approximately 498 additional units of absorption are required to give a reverberation time of 1.15 sec. Suppose, for example, that we select Fiberglas Acoustical Tile, type TXW, as an absorbing agent. From the table on page 33 we see that this material has an absorption coefficient of 0.69. Hence the total wall area to be covered with this material may be found by dividing 498 by 0.69, which gives 714 sq ft. In this particular room the ceiling has an area of 1250 sq ft. It is thus evident that slightly more than 58 per cent of the ceiling (or equivalent wall area) should be covered with the absorbing material in order to reduce the reverberation time to an acceptable value.

It is thus obvious that it is a comparatively simple matter to determine what remedial measures may be necessary in order to prevent the serious overlapping of successive sounds with the consequent deleterious effects on spoken words and the rendition of music. Nothwithstanding this fact many churches and other meeting places are designed and constructed without preliminary consideration being given to the acoustical problems involved. The results in such cases are disappointing, expensive, and sometimes impossible to correct. As these lines are being written the author has been asked to advise a certain church as to what, if anything, can be done to correct a serious reverberation problem in a new church building. The edifice has a beautiful interior and cost something like $350,000, but it is difficult for the congregation to understand what the minister says. A survey indicates that no steps were taken by the building committee, or the architect, to make provision for proper acoustical treatment. It is difficult to understand why those responsible for the construction of various public buildings will disregard the acoustical aspects of the undertaking, but such unfortunately is too often the case.

17–3. *Sound Insulation*

The acoustics of a room interior involves other factors in addition to the control of reverberation. For instance, the transmission of disturbing sounds into or from the room may be, and often is, a matter of concern. This is particularly true in connection with the design and construction of music studios and practice rooms. In such cases the reduction of sound transmission through the walls and other structural elements becomes important.

It is sometimes felt that if the interior surfaces of a room are acoustically treated in order to control reverberation the absorbing material used for that purpose will greatly decrease the transmission of unwanted noises through such walls. Unfortunately this is not true; the reduction will be only a few decibels—an insignificant amount.

If one is dealing with a solid masonry wall such as brick, concrete, or tile, the transmission loss depends only on the weight of the wall material per cubic foot. The heavier the wall the greater will be the reduction of sound transmission.

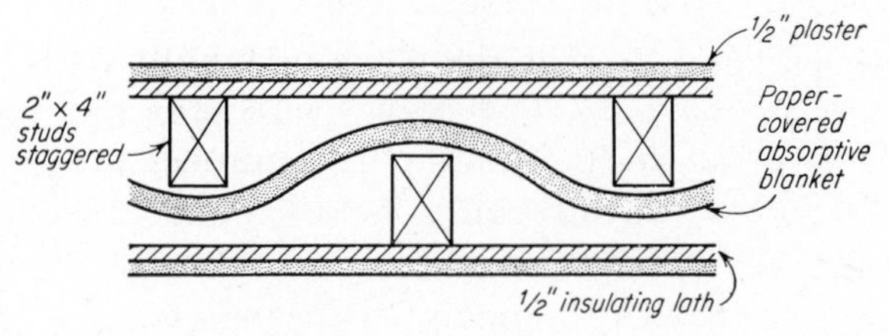

FIG. 17–6. One of several plans whereby the transmission of sound through walls may be reduced.

The effect of the massiveness of a wall structure holds, in a general way, for any type of material. Hence, because of its relatively light weight, a simple wooden partition will offer little resistance to the passage of sound waves. In fact, such a structure may even augment sound transmission. The wall may, for instance, be thrown into vibration by the sound waves from a source and thus act as a large resonator, somewhat like the sounding board of a piano. However by proper construction of a wall it is possible to obtain a transmission loss of as much as 50 db. An effective arrangement of component elements is sketched in Fig. 17–6. An absorption factor of at least 45 db is required in the walls of band rooms, music practice rooms, radio and sound studios.

In providing sound insulation for such rooms it is important to give attention to the construction of the doors and windows. Ordinary doors offer but small resistance to the passage of sound. However, there are available specially constructed doors that will, when properly installed, reduce the transmission by 40 db.

In the case of windows, such as those between a practice room and hall, or for observation windows in radio and recording studios, the construction should consist of at least three panes of heavy glass. The pieces of glass should be of different thickness: quarter-inch, three-eighths, or half-inch is recommended—this to avoid resonance effects. Each pane should be insulated completely from the frame by gaskets of rubber or heavy felt around all four edges. Further, the panes of glass should make a slight angle with each other, and not be closer than one inch from the next nearest pane.

17–4. *Reinforcement of Sound*

It sometimes happens that there are certain areas in an auditorium in which the sound intensity level is too low to produce satisfactory auditory results. Such places are usually on the main

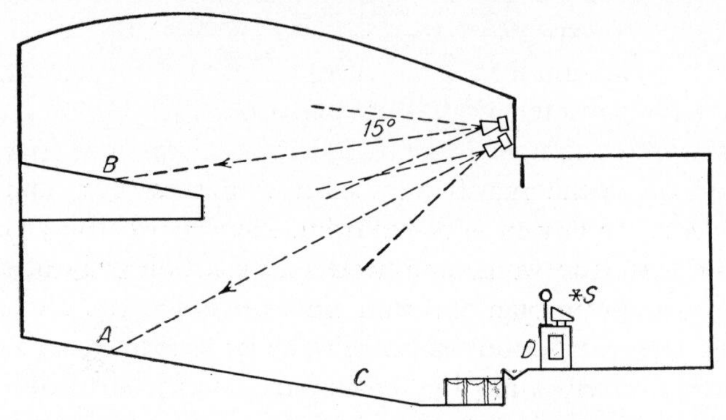

Fig. 17–7. Arrangement of loudspeakers and microphone whereby uniform sound distribution in an assembly room may be attained.

floor underneath the balcony and at the rear of the balcony. Modern electroacoustical facilities make it possible to correct such conditions, providing careful consideration is given to all factors involved. The method of applying remedial acoustical measures can best be outlined by considering the case of a typical auditorium in which it is desired to augment the original primary sound in the rear of the room. Suppose we have a situation such as that shown in the diagram of Fig. 17–7. This might represent a high school or chapel auditorium or any assembly hall. Let us assume that it is difficult to hear satisfactorily in the regions of *A* and *B* an average speaker stationed at *S*. Our problem is to augment the sound at *A* and *B* without doing so in the front of the auditorium, say in the region of *C*. In other words, we must aim to establish uniform sound intensity over the entire seating area.

To accomplish this end use is made of what is commonly referred to as a public-address system. Such a system, consisting of a microphone, an amplifier, and one or more loudspeakers, should be so engineered that the auditors are not conscious that a sound system is in operation, so far as fidelity and "presence" are concerned. Unfortunately many public-address systems do not meet such a standard. The chief reason for poor results is that second-rate components are used. Even for voice work, the best obtainable units should be utilized. The microphone should be a unidirectional model of the low-impedance electrodynamic type, and the amplifier should be a high-fidelity unit having twice the rated power that one expects to use.

The type and number of speakers to be used will depend upon the size and character of the audience room. For a church or other auditorium seating not to exceed 2000 people, two properly housed 12 in. coaxial speakers will yield satisfactory results. A plan, sometimes unwisely followed, of installing a multiplicity of small speakers along the two side walls is entirely unsatisfactory. Because of the difference in path length, the sound from the more remote units will reach the listener out of phase with the words arriving directly from the speaker, with the result that auditory confusion will obtain. The position of the two speakers is important. In a room of the size mentioned above, one should be installed on each side wall at not less than 20 ft nearer the audience than the microphone and at a height of 20 to 25 ft. The speakers must not be behind, or in the same vertical plane as, the microphone; otherwise acoustical "feedback" with the resulting "howling" will occur. The axis of both beams should be directed toward the front edge of the balcony. The position and direction of the speakers, as indicated above, will result in a uniform sound distribution throughout all of the seating area. Figure 17–8 shows a representative installation.

The two speakers may be operated at from 5 to 10 watts each, depending on the size of the room. If the public-address system is laid out in advance of building construction, the speaker units can usually be located so that they will be relatively inconspicuous.

FIG. 17–8. Section of interior of a college chapel. Note the location of the loudspeaker (*upper right*) with respect to the position of the microphone. Two such speakers were used, one on either side.

For larger auditoriums the reentrant type of horn speaker is to be preferred. If possible, a group of three or more such units should be mounted in a fan-shaped cluster, at or near the center of the proscenium arch.

When installing a sound system it is highly important that the units be properly phased; which means that all of the diaphragms shall move outward and inward simultaneously. Further, the total load impedance should match the output impedance of the amplifier.

It should be noted in passing that basically defective building acoustics cannot be corrected by the installation of a sound system, as is sometimes believed.

In out-of-door sound installations, horn transducers are usually employed. The amount of energy needed is considerably in excess of that required for an equal inside audience, the reasons being that there is little if any reflection, and extraneous noise may be present. However, the fundamental principles outlined above are applicable when planning for sound reinforcement in open spaces.

In concluding our discussion of sound reinforcement, it should be pointed out that each building constitutes a distinct problem, and before deciding upon the size and type of installation to be made it is important to make a preliminary acoustical survey of the particular auditorium involved. Such a survey consists of experimentally determining the intensity level at various points throughout the seating area. This involves the use of a constant source of sound, a microphone, and a suitable amplifier to the output of which is connected what amounts to an audibility meter or level indicator. A uniform level of not less than 60 db has been found to be satisfactory in public auditoriums.

The college chapel, before referred to, while free from serious reverberation, did, however, have an uncomfortably low sound level for auditors seated underneath the balcony and in rear balcony seats. This condition was remedied by the installation of a correctly designed 10-watt public-address system using a high-fidelity unidirectional microphone and two speakers. The relative position of the speaker units and the microphone can be seen in the illustration shown as Fig. 17–8.

The author is disposed to suggest to those readers who may chance to be responsible for the selection and installation of such equipment in connection with schools, colleges, or other auditoriums that it is advisable to consult a competent professional acoustician before making a final decision as to the purchase of equipment of this character. There are many persons calling themselves "sound engineers" who make promises of attaining satisfactory results at low cost; when, as a matter of fact, many such self-styled "sound experts" are not acquainted with the most elementary principles of acoustical engineering. A competent sound engineer will first make an acoustical survey, as outlined above, and then draft recommendations based on scientifically obtained data. To proceed on any other basis will lead to disappointment and needless expense.

17–5. *Cooperation Needed*

From what has been said in this and preceding chapters of this book, it is evident that closer cooperation between architects and acousticians is needed. By cooperative advance planning better acoustical results can be attained, and at less expense.

In concluding our brief study of musical acoustics the author also cannot refrain from appealing to musicians and the producers of musical instruments to make greater use of the tools which science has provided in this important field. Music is, indeed, the noblest of the arts. Its language is universal. It knows no limitations of race or creed or station. It brings joy to childhood; it lifts man's soul above the daily world of toil; it softens the sorrow of life's tragic hours. Because music means so much, to so many, every possible effort should be made to enhance its beauty and its usefulness. Through the sympathetic cooperation of the scientist and the musician, music can be made a more effective agent for the enrichment of man's intellectual and spiritual life.

Bibliography

Barbour, J. M.: "Tuning and Temperament," Michigan State College Press, East Lansing, Mich., 1951.

Barnes, W. H.: "The Contemporary American Organ," 5th ed., J. Fischer & Brother, New York, 1952.

Bartholomew, W. T.: "Acoustics of Music," Prentice-Hall, Inc., New York, 1942.

Barton, E. H.: "Textbook of Sound," Macmillan & Co., Ltd., London, 1908.

Beranek, L. L.: "Acoustical Measurements," John Wiley & Sons, Inc., New York, 1949.

Buck, P. C.: "Acoustics for Musicians," Oxford University Press, New York, 1918.

Colby, M. Y.: "Sound Waves and Acoustics," Henry Holt and Company, Inc., New York, 1938.

Coleman, S. N.: "Bells: Their History, Legends, Making, and Uses," Rand McNally & Company, Chicago, 1928.

Crandall, I. B.: "Theory of Vibrating Systems and Sound," D. Van Nostrand Company, Inc., New York, 1920.

Dorf, R. H.: "Electronic Musical Instruments," Radio Magazines, Inc., Mineola, New York, 1954.

Duhamel, G.: "La Flûte," Libraire Gründ, Paris, 1953.

Eby, R. L.: "Electronic Organs," Van Kampen Press, Wheaton, Illinois, 1953.

Fletcher, H.: "Speech and Hearing," D. Van Nostrand Company, Inc., New York, 1929.

Hamilton, C. G.: "Sound and Its Relation to Music," Oliver Diston, Inc., Boston, 1912.

Hart, G.: "The Violin: Its Famous Makers and Their Imitators," Dulau & Co., London, 1884.

Helmholtz, H. von: "Sensations of Tone," trans. A. J. Ellis, Longmans, Green & Co., Inc., New York, 1912.

Hipkins, A. J.: "Musical Instruments," A. & C. Black, Ltd., London, 1921.

Jeans, J.: "Science and Music," The Macmillan Company, New York, 1937.

Jones, A. T.: "Sound," D. Van Nostrand Company, Inc., New York, 1937.

Kinsler, L. E.: "Fundamentals of Acoustics," John Wiley & Sons, Inc., New York, 1950.

Knudsen, V. O.: "Architectural Acoustics," John Wiley & Sons, Inc., New York, 1932.

——— and C. M. Harris: "Acoustical Designing in Architecture," John Wiley & Sons, Inc., New York, 1950.

Koenig, R.: "Quelques expériences d'acoustique," A. Lahure, Paris, 1882.

Kostelijk, P. J.: "Theories of Hearing," Universitaire Pers Leiden, Leiden, 1950.

Loyds, L. S.: "Music and Sound," Oxford University Press, New York, 1937.

Mersenne, Père: "Harmonie universelle," 1636.

Miller, D. C.: "The Science of Musical Sounds," The Macmillan Company, New York, 1916.

Mills, J.: "A Fugue in Cycles and Bels," D. Van Nostrand Company, Inc., New York, 1935.

Morse, P. M.: "Vibration and Sound," 2d ed., McGraw-Hill Book Company, Inc., New York, 1948.

Olson, H. F.: "Elements of Acoustical Engineering," D. Van Nostrand Company, Inc., New York, 1940.

———: "Musical Engineering," McGraw-Hill Book Company, Inc., New York, 1952.

Rayleigh, J. W. S.: "Theory of Sound," Dover Publications, New York, 1945.

Rensch, R.: "The Harp," Philosophical Library, Inc., New York, 1950.

Richardson, E. G.: "The Acoustics of Orchestral Instruments and of the Organ," Edward Arnold & Co., London, 1929.

Sabine, W. C.: "Collected Papers in Acoustics," Harvard University Press, Cambridge, Mass., 1922.

Smith, H.: "Sound in the Organ and Orchestra," Charles Scribner's Sons, New York, 1911.

Stanley, D., and J. P. Maxfield: "The Voice: Its Production and Reproduction," Pitman Publishing Corporation, New York, 1933.

Stevens, S. S., and H. Davis: "Hearing," John Wiley & Sons, Inc., New York, 1938.

Stewart, G. W.: "Introductory Acoustics," D. Van Nostrand Company, New York, 1933.

——— and R. B. Lindsay: "Acoustics," D. Van Nostrand Company, Inc., New York, 1930.

Taylor, S.: "Sound and Music," Macmillan & Co., Ltd., London, 1873.

Tyndall, J.: "Sound," Longmans, Green & Co., Ltd., London, 1867.

Watson, R. F.: "Acoustics of Buildings," John Wiley & Sons, Inc., New York, 1932.

Wever, E. G.: "Theory of Hearing," John Wiley & Sons, Inc., New York, 1949.

White, W. B.: "Piano Tuning and Allied Arts," Tuners Supply Co., Boston, 1938.

Whitworth, R.: "The Electric Organ," Musical Opinion, London, 1930.

Wood, A. B.: "A Textbook of Sound," George Bell & Sons, Ltd., London, 1930.

Zahm, J. A.: "Sound and Music," A. C. McClurg & Co., Chicago, 1892

Index